**CTBUH** Technical Guide

# Natural Ventilation in High-Rise Office Buildings

An output of the CTBUH Sustainability Working Group

Antony Wood & Ruba Salib

NEW YORK AND LONDON

Principal Authors: Antony Wood & Ruba Salib
Coordinating Editor & Design: Steven Henry
Editing Support: Candace Baitz

First published 2013 by Routledge
2 Park Square, Milton Park, Abingdon, Oxon, OX14 4RN

Simultaneously published in the USA and Canada by Routledge
711 Third Avenue, New York, NY 10017

*Routledge is an imprint of the Taylor & Francis Group, an informa business*

Published in conjunction with the Council on Tall Buildings and Urban
Habitat (CTBUH) and the Illinois Institute of Technology

*British Library Cataloguing in Publication Data*
A catalogue record for this book is available from the British Library

*Library of Congress Cataloging-in-Publication Data*
A catalog record has been requested for this book

ISBN13 978-0-415-50958-9

Council on Tall Buildings and Urban Habitat
S.R. Crown Hall
Illinois Institute of Technology
3360 South State Street
Chicago, IL 60616
Phone: +1 (312) 567-3487
Fax: +1 (312) 567-3820
Email: info@ctbuh.org
http://www.ctbuh.org

*Front Cover: KfW Westarkade, Frankfurt Germany (see pages 122–131). © Sauerbruch Hutton / photo: Jan Bitter*

## Principal Authors

*Antony Wood,* Council on Tall Buildings and Urban Habitat
*Ruba Salib,* University of Nottingham


## Contributors/Peer Review

*Alan Barbour,* Beca Group
*Seifu Bekele,* Vipac
*Robert Bolin,* Syska Hennessy Group
*Gail Brager,* University of California at Berkeley
*Steven Cook,* Murphy/Jahn Architects
*Brian Ford,* University of Nottingham
*Joana Carla Soares Gonçalves,* University of São Paulo
& Architectural Association School of Architecture
*Richard Hassell,* WOHA Architects
*Christoph Ingenhoven,* ingenhoven architects
*Dr. Shireen Jahnkassim,* International Islamic University Malaysia
*Mehdi Jalayerian,* Environmental Systems Design, Inc.
*Phil Jones,* Cardiff University
*Nico Kienzl,* Atelier Ten
*Luke Leung,* Skidmore, Owings & Merrill LLP
*Ian Maddocks,* Buro Happold
*Erin McConahey,* Arup
*Hannah Morton,* Cundall
*David Nelson,* Foster + Partners
*Daniel O'Connor,* Aon Fire Protection Engineering
*Tomokazu Ohno,* Meiji University
*Geoffrey Palmer,* Grontmij Ltd.
*Lester E. Partridge & Simon Lay,* AECOM
*Mark Pauls,* Manitoba Hydro
*John Peterson,* KPMB Architects
*Duncan Phillips,* RWDI
*Carme Pinós,* Estudio Carme Pinós
*Gary Pomerantz,* WSP Flack + Kurtz
*Kevin Powell,* U. S. General Services Administration
*John Pulley,* HOK
*Matthias Sauerbruch & Louisa Hutton,* Sauerbruch Hutton
*Matthias Schuler, Alex Knirsch & Stefi Reuss,* Transsolar
*Wolfgang Sundermann,* Werner Sobek Group
*Peter Weismantle,* Adrian Smith + Gordon Gill Architecture
*Dr. Ken Yeang,* T. R. Hamzah & Yeang Sdn. Bhd. & Llewelyn Davies Yeang

# Contents

# About the CTBUH

The Council on Tall Buildings and Urban Habitat is the world's leading resource for professionals focused on the design and construction of tall buildings and future cities. A not-for-profit organization based at the Illinois Institute of Technology, the group facilitates the exchange of the latest knowledge available on tall buildings around the world through events, publications, and its extensive network of international representatives. Its free database on tall buildings, The Skyscraper Center, is updated daily with detailed information, images, and news. The CTBUH also developed the international standards for measuring tall building height and is recognized as the arbiter for bestowing such designations as "The World's Tallest Building."

# About the Authors

### Antony Wood
*Council on Tall Buildings and Urban Habitat*

Antony Wood has been Executive Director of the Council on Tall Buildings and Urban Habitat since 2006. He is chair of the CTBUH Tall Buildings and Sustainability Working Group. Based at the Illinois Institute of Technology, Antony is also an Associate Professor in the College of Architecture, where he convenes various tall building design studios. A UK architect by training, his field of specialism is the design, and in particular the sustainable design, of tall buildings. Prior to becoming an academic, Antony worked as an architect in practice in Hong Kong, Bangkok, Kuala Lumpur, Jakarta, and London. He is the author and editor of numerous books and papers in the field, including the 2008 title "Tall & Green: Typology for a Sustainable Urban Future." His PhD explored the multi-disciplinary aspects of skybridge connections between tall buildings.

### Ruba Salib
*University of Nottingham*

Ruba was raised in Amman/Jordan and immigrated to Canada in 2002, where she earned her Bachelor degree from the University of Toronto, majoring in both architecture and fine art history and developing an interest in sustainable design and energy efficient buildings. After practicing for a year at an architecture firm in Jordan, Ruba pursued her Master of Architecture in Environmental Design at the University of Nottingham in the UK, with a dissertation topic on Natural Ventilation in High-Rise Office Buildings. A passion for sustainable design has led her to be actively involved in drafting green building guidelines for Jordan and the United Arab Emirates. Currently based in London, she is working on various small-scale and large-scale international projects at RSP Planet Design Studios, an architecture, planning and engineering practice headquartered in Singapore.

# Preface

Increasing density in cities is now widely accepted as necessary for achieving more sustainable patterns of life to reduce energy consumption and thus combat climate change. The concentration of people in denser cities – sharing space, infrastructure, and facilities – offers much greater energy efficiency than the expanded horizontal city, which requires more land usage as well as higher energy expenditure in infrastructure and mobility. This is especially true in the face of the two major population shifts the human race is currently experiencing – rapid population growth and massive urbanization.

The United Nations forecasts 70 percent of the world's projected nine billion population will be urbanized by the year 2050, up from 51 percent of seven billion urbanized as of 2010.[1] The impact of this total figure of 2.8 billion people moving into cities over the next 40 years is perhaps better understood at the annual rate of 70 million people per year, or the daily rate of nearly 200,000 people. As a global species it means we need to build a new or expanded city of more than one million people every week for the next 40 years to cope with this urban growth. In the context of these numbers, it is clear that a continued horizontal spread of cities is unsustainable. Our urban agglomerations need to become much denser.

Tall buildings are not the only solution for achieving density in cities but, given the scale of these major population shifts, the vertical city is increasingly being seen as the most viable solution for many urban centers – especially those in developing countries such as China or India where population growth and urbanization is at its most pronounced. However, the full implications of concentrating more people on smaller plots of land by building vertically, whether for work, residential or leisure functions, needs to be better researched and understood.

---

[1] Source: Population Division of the Department of Economic and Social Affairs of the United Nations Secretariat, World Population Prospects: The 2006 Revision and World Urbanization Prospects: The 2007 Revision, http://esa.un.org/unup

On the one hand there are many energy benefits from building tall. In addition to the larger-scale benefits of density versus horizontal spread, tall buildings offer advantages such as less materials (and thus reduced carbon) needed for enclosure per square foot of usable floor space created, a smaller surface area of envelope per floor area for heat loss/gain, a natural energy share that occurs between floors (especially heat energy in colder climates), plus the potential for harvesting solar and wind energy at height.

On the other hand there are disadvantages with building tall that offset, and may even negate, the benefits of concentrating people together in taller buildings. A smaller surface area per floor area may limit contact between occupier and envelope, affecting access to natural light, view and ventilation, and possibly leading to a lower quality of internal environment. Materials at height – whether primary structural systems to counteract wind loads or curtain wall systems needing to counteract greater environmental pressures – require greater sizing and performance than in the low-rise realm, further affecting the overall sustainability equation. There are also a host of other factors – societal as well as operational – which are not yet fully investigated, nor likely maximized in their potential.

The general concept of "vertical" being more sustainable than "horizontal" may thus be true, especially when the larger-scale urban scenario is considered, but the myriad factors that contribute to this scenario need to be better researched and understood. Building owners, developers and consultants need to be able to understand the "sustainability threshold" for height – that height or floor count figure beyond which additional height would not make sense on sustainable grounds (and likely cost grounds as well). Of course this figure will never be an exact science and will differ not only from city to city, but from site to site and building to building. It is, however, a measure of extreme importance – and one which the global building industry needs to urgently strive toward.

### Objectives of this Guide

The CTBUH Tall Buildings and Sustainability Working Group has set out to determine this "sustainability threshold" for height, despite the complex and varying nature of the equation, initially through a series of guides that analyze each aspect of the tall building in turn. Often there is disagreement and debate on what constitutes the most sustainable principles and systems for a tall building – indeed any building – but, irrespective of the final solutions, it is generally accepted that we need to reduce the energy equation – in both operating and embodied terms – of every component and system in the building as part of making the entire building more sustainable. Each guide in this expected series then seeks to establish best-practice and beyond for each system.

Though the exact percentage will vary based on climate, the HVAC (heating, ventilation and air-conditioning) systems in tall office buildings[2] typically account for 33 percent or more of overall building energy consumption (see Figure 1). Of that percentage, more than half of the HVAC energy is due to overcoming the heat gains due to occupants, lighting, miscellaneous power

**Given that the HVAC systems in tall office buildings typically account for somewhere between 30–40 percent of overall building energy consumption, the elimination of these systems with natural ventilation could be argued to be the most important single step we could take in making tall buildings more sustainable.**

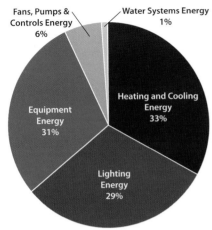

▲ Figure 1: Average energy consumption of a typical high-rise office building. (Source: Interpreted from the US Department of Energy's reference building energy models for existing large commercial buildings built after 1980, across 16 US cities, in various climates.)

(e.g., plug loads) and solar and thermal gains and losses. The increased efficiency – or possibly even elimination – of these systems could thus be argued to be the most important single step in making tall buildings more sustainable. The reduction of this reliance on "mechanical" ventilation – through the "re-introduction" of natural ventilation systems in our buildings – is thus the focus of this guide.

It should be noted that there are numerous tall office buildings in existence (many of which are profiled in the case studies section of this report) which employ innovative natural ventilation systems, often for the greater part of the operating year. It is extremely rare, however, for a significant tall office building to be able to rely 100 percent on natural ventilation[3] – due largely to the implications of failure of the system – and this fact alone is a strong driving force for this guide. Even the best case buildings employ "hybrid" systems for ventilation (i.e., typically using natural ventilation for periods when the external conditions allow, but then full mechanical systems take over when external conditions are not optimal due to temperature, humidity, noise or pollution). Referred to as "Mixed-Mode," these hybrid systems result in significant energy

[2] While the natural ventilation of *all* buildings is beneficial, it is generally accepted that natural ventilation in tall *office* buildings – with greater floor area sizes/depths, higher population density and higher internal heat gains through equipment, among other factors – is more difficult to achieve than in residential or hotel buildings with smaller floor plate depths, cellular "room" layouts, and greater contact with the outside envelope. This guide thus concentrates on the more difficult task of naturally ventilating tall office buildings, though many of the findings and principles could be adopted in tall residential buildings, or tall buildings of other functions.

[3] There is only one completed building included as a case study in this guide – the Torre Cube, Guadalajara – which is fully naturally ventilated without any mechanical plant for heating, cooling or ventilation. The building, however, is only 17 stories tall and takes great advantage of the compliant Guadalajara climate. The transferability of its systems and approaches to other locations would thus be difficult at best.

savings over the year when natural ventilation is enabled instead of mechanical systems; however, the true potential of natural ventilation is not delivered since a mechanical plant needs to be provided to cope with the peak load (worst case scenario) conditions. The embodied energy required for the mechanical plant equipment, not to mention the loss of floor space and impact on other aspects of the building, can be significant. The ideal scenario then – and one which this guide strives toward – is for tall buildings which deliver acceptable internal conditions throughout the whole year through natural ventilation alone.

## Content Overview

The term "re-introduction" of natural ventilation systems is used intentionally to remind us that, for many centuries, buildings relied on natural ventilation strategies alone. In the post-World War II period when "cheap" energy liberated architecture from its connection with the natural environment, the advent of the sealed, air-conditioned Modernist box proliferated around the world. Prior to that period, all buildings were naturally ventilated as a matter of necessity since mechanical ventilation systems were not yet advanced enough to sufficiently condition the space. Further, the systems to achieve natural ventilation differed from place to place according to climate and, sometimes, culture. Ultimately, much of what this guide proposes – buried though it may be beneath layers of modern solutions and technical systems – is a re-learning of the principles upon which the ventilation of buildings were based for many hundreds of years.

It is perhaps fitting then that a brief historical overview of ventilation in tall office buildings forms the first section to this guide, followed by a brief outline of the generic principles of natural ventilation as a background to the more advanced systems outlined later. The majority of the guide is focused on the analysis of a number of seminal case studies that employ natural ventilation systems to a greater or lesser degree – buildings ranging in completion from 1996 to 2011, and locations from Frankfurt to Guadalajara. Fifty percent of the case studies are located in Germany. This is the result of the relatively moderate climate of the region, which allows for natural ventilation systems to be implemented, combined with Germany's generally ambitious commitment to the development of these systems.

Before embarking on the following pages, the reader should note that we began the compilation of this guide fully intending to deliver detailed performance data on each case study, which would allow a direct analysis of the effectiveness of each system and strategy employed, for the benefit of those considering such strategies in other buildings. That we have fallen somewhat short of this high ideal reflects, in the authors' opinion, perhaps the single biggest barrier facing the industry in the true adoption of sustainable design principles moving forward, i.e., the lack of hard data verifying what is working effectively, and what isn't.

Though there have been moves in recent years to address this – sustainability ratings systems insisting on energy reporting, for example – the reality is that we are still a long way from the open sharing of reliable information on building energy performance, or even the agreed metrics to allow a fair and accurate comparison across buildings and building types. Even where the data does exist there is still a general reluctance to release it into the public domain – especially for high-profile buildings. Some of the case studies in this guide were completed relatively recently, complicating any attempt to build up reliable and consistent data within the time frame of this publication. Other buildings included, however, have been in operation for a decade or more, and yet the data still was not available.

Where any information on building performance was obtained, it has been included to give the reader the best possible understanding and assessment of the systems employed. However, it should be noted that, in many cases, even this information is estimated based on predictive analysis studies that were undertaken at the design stage.

A qualitative analysis of the strengths and limitations of each system – often developed in conjunction with the consultants connected with each case study – has been included. A comparative analysis has also been drawn between the different case studies in order to determine the key parameters which influence the prospects of natural ventilation in high-rise office buildings. This analysis is drawn together into a set of considerations and recommendations which conclude this guide, together with suggested areas for further necessary research in the field.

The guide is aimed primarily at the typical building owner or professional who wants a better understanding of the options available for naturally ventilating a tall office building. It does not pretend to portray any radical new knowledge that previously did not exist. Rather, it aims to bring together disparate strands of information and put them together, with analysis, into one publication. This is intended as the first of several guides from the CTBUH Tall Buildings & Sustainability Working Group that, when taken together, will provide a tool kit for the creation of tall buildings with a much-reduced environmental impact, while taking the industry closer to an appreciation of what constitutes a sustainable tall building, and what factors affect the "sustainability threshold" for tall buildings.

**Even the best case buildings employ "hybrid" systems for ventilation, typically using natural ventilation for periods when the external conditions allow, but then full mechanical systems take over when external conditions are not optimal.**

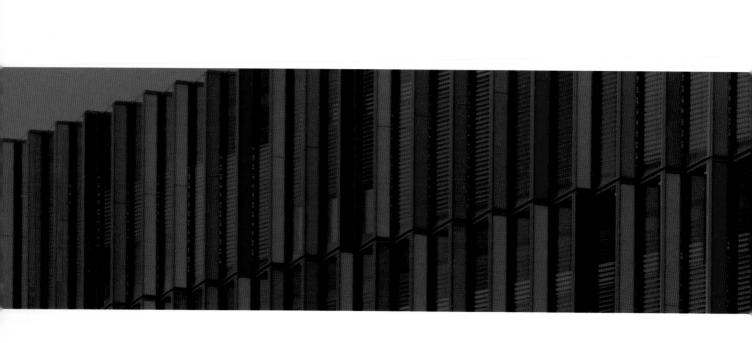

# 1.0 Introduction and Background

# 1.0 Introduction and Background

## 1.1 Historical Overview of Natural Ventilation in High-Rise Office Buildings

The 19th century marked the emergence of the office building typology as we know it today. Controlling the indoor environment in these early office buildings was achieved by passive means. In most US and European cities, operable windows were used for natural ventilation and for keeping cool, while stoves and radiators were the main sources of heat energy when it was cold. The high cost of electricity and the need to conduct tasks under natural lighting conditions had a profound impact on the design of office buildings. The need to provide adequate daylight limited the depth of office floor plans, and consequently enabled natural ventilation by means of operable windows.

Although keeping cool was not a major concern for architects at the time, natural ventilation was considered necessary for sanitary purposes and for the elimination of excessive humidity. Furthermore, many of the large office buildings during that period were influenced by the classical styles of architecture, which involved the use of central open courts, or light-wells, that limited plan depths to allow natural light and air into the interior.

This building type became particularly common in Chicago during the office building boom that followed the Great Fire of 1871. The building type was referred to as the "Chicago Quarter Block" because it employed open courts and occupied the plot of the entire city block between streets. This building type was also exported to many other US cities, where it was used as a model for emerging office buildings.

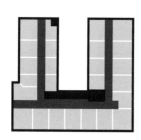

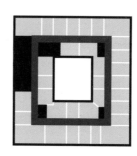

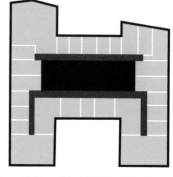

▲ Figure 1.1: View of the 1891 Wainwright Building, St. Louis (top); and typical floor plan (bottom). © Antony Wood/CTBUH

▲ Figure 1.2: View of the 1924 Straus Building, Chicago (top); and typical floor plan (bottom). © Marshall Gerometta/CTBUH

▲ Figure 1.3: View of the 1930 Chrysler Building, New York (top); and typical floor plan (bottom). © Steven Henry/CTBUH

Light courts were integrated into the buildings in E, H, and U-shaped plan arrangements (Arnold 1999a, pp. 40–54). Louis Sullivan's Wainwright Building in St. Louis, which was completed in 1891, is a good example of the adoption of this classical style (based on the Uffizi in Florence) through the use of a U-shaped plan to provide light and air to every office (see Figure 1.1). It is also worth noting that external sunshades were integrated into the design of the Wainwright Building as a passive means of controlling solar gains and providing thermal comfort.

Built in 1924, the Straus Building in Chicago (a 21-story building with a nine-story tower) set another example of how the classical style was adopted through the use of a large central light court (see Figure 1.2).

At the beginning of the 20th century, architects started to design many of the most prominent skyscrapers in New York on the same basis. Iconic buildings such as the Chrysler Building (1930), and the Empire State Building (1931) reached unprecedented heights while still relying on natural ventilation and lighting. The form of these skyscrapers and the depth of their plans (see Figure 1.3) were still driven by the need to provide natural light for office interiors, with no particular emphasis given to the development of a natural ventilation strategy.

By the 1950s, the availability of cheap energy and the widespread use of air-conditioning had a profound impact on the form and planning of office buildings. The ability to control indoor temperature and humidity by mechanical means eliminated the restrictions architects faced with regards to plan form, plan depth, and window fenestrations. In other words, the consideration of passive measures to provide comfortable indoor environments were no

longer a central concern for architects and engineers at that time. The dependence on air-conditioning allowed the emergence of deep-planned, transparent office buildings with curtain-walled windows. The heavyweight stone or brick-clad skyscrapers of the early 1900s were replaced by the light, fully-glazed office buildings of the 1950s and 1960s. Renowned architect Mies van der Rohe's work in high-rise buildings exemplifies this era of the all-glass-box style of architecture, from his 1958 Seagram Building in New York (see Figure 1.4), to his 1972 IBM Building in Chicago. The increased transparency and lightness of the structure, as well as the lack of solar shading devices, placed a higher load on air-conditioning systems to cool down buildings during the summer and to heat them during the winter (in temperate climates).

The oil crisis of 1973 marked another turning point in the development of office buildings. As a result of this crisis, western countries aimed to reduce global energy consumption – mainly by reducing the energy used in buildings for heating, cooling, and ventilation. The proposed solutions mainly focused on increasing the insulation level of building envelopes and reducing the air infiltration level by sealing the building. In other words, the main goal was to reduce the consumption of fuel by reducing heat loss through ventilation. The introduction of these increasingly sealed office buildings impacted negatively the comfort and health of the occupants. This resulted in the deterioration of indoor air quality and the spread of diseases due to humidity condensation and the growth of mold (e.g., "sick building syndrome"). The lack of fresh air and the overheating problems in summer affected the productivity and performance of office workers.

As a response to the oil crisis and the development of building-related illnesses, designers started to consider energy conservation measures that provided healthier and more comfortable working environments. The 1980s and 1990s thus marked the start of greater building energy efficiency and a return to considering the benefits of natural ventilation in buildings, as well as passive heating and cooling strategies in office building designs. This report will focus on the most advanced strategies adopted to naturally ventilate high-rise office buildings during the late 20th and early 21st centuries.

▲ Figure 1.4: View of the 1958 Seagram Building, New York (top); and typical floor plan (bottom). © Antony Wood/CTBUH

## 1.2 The Principles of Natural Ventilation in a High-Rise Building

The driving forces for natural ventilation in a tall building are of course the same as those for other buildings. The physical mechanisms for natural ventilation rely on the pressure differences generated across the envelope openings of a building. The pressure differences are generated by:
(i) the effects of wind,
(ii) temperature differences (gravity acting on density) between inlet and outlet of air, or
(iii) a combination of both.

Natural ventilation can therefore be categorized into "wind-induced" and "buoyancy-induced" ventilation according to the physical mechanism driving the air. Wind-induced ventilation occurs when wind creates a pressure distribution around a building with respect to the atmospheric pressure (see Figure 1.5). The pressure differences drive air into the building's envelope on the windward side (positive pressure zone) and out of the building through the openings on the leeward side (negative pressure zone). The pressure effect of the wind on a building is primarily dominated by the building's shape, the wind direction and velocity,

and the influence of the surroundings, which are all factors that influence the pressure coefficient. In addition to the value of the pressure coefficient, the mean pressure difference across a building's envelope is dependent upon the mean wind velocity at upwind building height, and the indoor air density as a function of atmospheric pressure, temperature, and humidity.

Buoyancy-induced ventilation (also known as "stack effect" or the "chimney effect") occurs due to density differences caused by variations in temperature and height between the inside and the outside or between certain zones within a building. The pressure differences generated by buoyancy are mainly dependant on the stack height (the height difference between air intake and extract openings) and the air density difference as a function of temperature and moisture content in the air. To guarantee inward airflow in the absence of wind, it is important to ensure that outdoor temperatures are lower than indoor temperatures to achieve buoyancy-induced ventilation. When the indoor air temperature exceeds the outdoor temperature, an under-pressure is formed in the lower part of a building, pulling air inwards through the openings in the envelope

(although, if outdoor temperatures are equal to indoor temperatures, buoyancy can still occur due to internal loadings). As the air travels through the building, it is heated by the internal gains and building occupants.

The density difference caused by the indoor/outdoor temperature difference results in a different pressure gradient in the building. The over-pressured zones at the top of the building drive air out of the openings in the building (since air flows from areas of high pressure to areas of low pressure). At a certain height of the building, however, the indoor pressure and the outdoor pressure are equal to each other. This level is referred to as the "neutral plane" or "neutral pressure level" (see

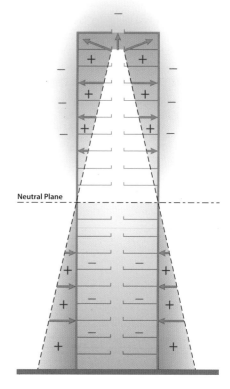

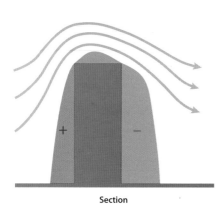

**Section**

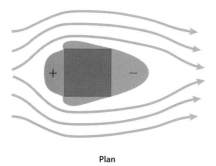

**Plan**

▲ Figure 1.5: Pressure distribution around a building due to wind flow.

▲ Figure 1.6: Thermal buoyancy in tall buildings where indoor/outdoor temperature differences result in a different pressure gradient in the building.

Figure 1.6). In order to achieve effective buoyancy-induced ventilation, there has to be a significant temperature differential between the inlet and outlet of air, and minimal internal resistance to air movement within the interior spaces (e.g., on most buildings where floor plates cover the whole floor area without significant voids or atria there is no significant pressure buildup over the height of the building). It should be noted that a "reverse stack effect" can occur when outside air temperatures are significantly higher than internal building temperatures. Under this condition air can enter high-rise buildings at high elevations and discharge at lower elevations. This reverse stack effect can be difficult to manage.

Finally, it is important to note that the two driving forces for natural ventilation (wind and buoyancy) can occur separately, but are more likely to occur at the same time. Thermal buoyancy will generally be the dominating driving force on a calm day with practically no wind, whereas pressure differentials generated by wind will typically be the dominating driving force on a windy day.

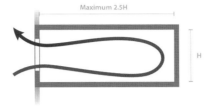

▲ Figure 1.7: Single-sided ventilation.

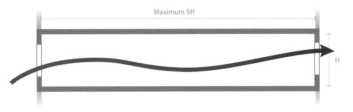

▲ Figure 1.8: Cross-ventilation.

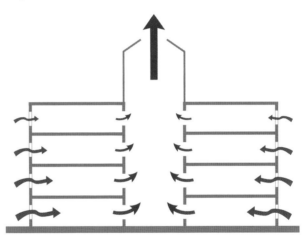

▲ Figure 1.9: Stack-ventilation.

## 1.3 Natural Ventilation Strategies

A "ventilation strategy" refers to how air is introduced into a building, and how it is extracted out of it. As in low-rise buildings, the different strategies used to ventilate high-rise buildings can be classified into three main categories:

*(i) Single-sided ventilation,* where fresh air enters the room through the opening on the same side it is exhausted from. This strategy can ventilate the space effectively if the room depth is a maximum of 2.5 times its height (see Figure 1.7). The driving force for single-sided ventilation is wind coinciding with the temperature difference between low-level air inlets and high-level air outlets. Buoyancy effect can also aid single-sided ventilation if the ventilation openings are located at different heights.

*(ii) Cross-ventilation,* which relies on the flow of air between the two sides of a building's envelope due to the pressure differentials between openings in the two sides (air moves from the windward to the leeward side). For effective cross-ventilation, the depth of the room must not exceed five times its height (see Figure 1.8). The buoyancy effect can also aid the effectiveness of cross-ventilation when the spaces are facing a tall open space such as an atrium.

*(iii) Stack-ventilation,* which involves the entry of fresh air into the building at a low level and its exhaust at a high level due to the occurrence of temperature, density, and pressure differences between the interior and exterior or between certain zones within a building. Stack-ventilation is often used in buildings which have a central atrium, chimney, or elevated part (see Figure 1.9).

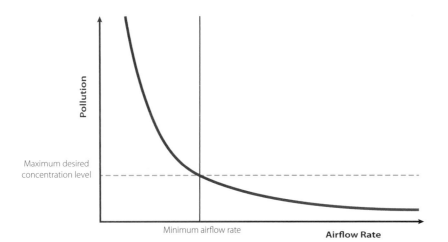

▲ Figure 1.10: Natural ventilation for indoor air quality showing the relation of pollution level to the airflow rate. (Source: Allard & Santamouris 1998, p. 3)

| Air Speed | Impact on Building Occupants |
|---|---|
| <0.25 m/s | Unnoticed |
| 0.25–0.5 m/s | Pleasant |
| 0.5–1.0 m/s | Awareness of Air Movement |
| 1.0–1.5 m/s | Drafty |
| >1.5 m/s | Annoyingly Drafty |

Note: Higher velocities are acceptable in hotter and humid climates

▲ Figure 1.11: Table showing the impact of various air speeds on occupant comfort levels in a building. (Source: Auliciems & Szokolay 1997, p. 14)

### Mixed-Mode Ventilation Strategies

The prospect of relying solely on natural ventilative cooling is limited by the local site and climatic conditions. For instance, the use of natural ventilation may not be appropriate for buildings in extreme climates (extremely cold, hot, and/or humid) or on sites with high levels of noise and pollution. In this case, mixed-mode or "hybrid" ventilation presents a suitable strategy that utilizes a combination of natural ventilation (from operable windows and/or vents) and mechanical systems which offer some form of cooling and air distribution. Mixed-mode buildings are generally classified on the basis of their operation strategies, which illustrate whether natural ventilation and mechanical cooling are operating in the same or different spaces in a building, and/or at the same or different times. As such, mixed-mode ventilation strategies are typically classified into the following categories:

**(i) Contingency,** where the building is designed either as an air-conditioned building with provision to switch to natural ventilation, or as a naturally ventilated building with space and electrical infrastructure allocated for the possible installation of mechanical equipment for air-conditioning in the future.

**(ii) Zoned,** where mechanical cooling and natural ventilation operate in different areas or zones of the building.

**(iii) Complementary,** where the building is designed with the capability to operate under mechanical and natural ventilation modes in the same space. This category is further subdivided into Alternate, Changeover, and Concurrent mixed-mode strategies as described below:

▸ *(a) Alternate,* where the building includes provisions and equipment for both air-conditioning and natural ventilation, but operates continuously in one mode or the other.

▸ *(b) Changeover,* where the building switches between mechanical cooling and natural ventilation on a seasonal or daily basis, depending on the outside weather conditions.

▸ *(c) Concurrent,* where mechanical cooling and natural ventilation can operate in the same space at the same time (e.g., when an HVAC system supplements natural ventilation to extend occupant comfort and maximize energy efficiency) (Brager et al. 2000, pp. 60–70).

Mixed-mode buildings often incorporate sophisticated building management systems and control strategies which allow for the overlap/alternation between the natural ventilation and mechanical cooling systems. Examples of various mixed-mode buildings will be discussed in further detail in the Case Studies section of this report.

### 1.4 The Purpose and Benefits of Natural Ventilation

Apart from improving the energy performance of a building, natural ventilation plays a key role in providing both good indoor air quality and acceptable thermal comfort conditions for occupants, if integrated correctly. In addition, the employment of natural ventilation in an office environment can help to mitigate noise and health problems (i.e., "sick building syndrome") associated with mechanical HVAC systems, providing a healthier and more comfortable environment for occupants and enhancing their productivity. Several studies on naturally ventilated and mixed-mode buildings (which utilize a combination of natural ventilation and mechanical HVAC systems) show examples of productivity improvements of 3–18 percent annually (Loftness, Hartkopf, & Gurtekin 2004). Based on an annual salary of US$45,000 this would be equivalent to annual productivity gains of US$3,900 per employee. Furthermore, the use of natural ventilation potentially reduces the embodied energy, capital, maintenance, and operational costs of mechanical equipment and the space needed to accommodate the equipment.[1] In high-rise buildings, HVAC systems can take up several floors to house mechanical equipment, depending on the building's height.

Optimum indoor air quality is defined as the supply of fresh air and the removal or dilution of indoor pollution. Typical indoor air pollutants that HVAC systems can be a conveyor of include:
(i) odor and moisture from humans and human activities,
(ii) off-gassing from building materials, furnishing, fittings, equipment, and cleaning products,
(iii) emissions from gas combusting appliances (such as $CO_2$, CO, $NO_2$ and particulate matter), and
(iv) radon and pollution from outdoor sources.

Figure 1.10 shows that the level of indoor pollution decreases exponentially as the natural ventilation airflow rate increases, and that a minimum airflow rate is required to prevent indoor pollution from exceeding the maximum permitted concentration levels. The velocity of air movement in office spaces should not be too high as it could cause loose papers to fly about and negatively affect occupants' thermal comfort levels (see Figure 1.11). It is important to find a balance between optimum indoor air quality, ventilation effectiveness, energy use, and thermal comfort. Comfort zones are usually between 18 °C and 24 °C with a relative humidity between 30–60 percent (Yeang 2006). Yet, human thermal comfort and tolerance can vary significantly according to cultural and other factors. For further specific information, see Section 3.1 Thermal Comfort Standards.

There are three ways in which natural ventilation can improve human comfort:
(i) by cooling indoor air through incorporating outdoor air as long as external temperatures are lower than internal temperatures,
(ii) by cooling the building fabric,
(iii) by cooling the occupant directly through convection and evaporation (psychological cooling). Even in high humidity levels and air temperatures, the direct physiological cooling effect of ventilation passing over the skin of the body can be achieved with the introduction of air at a relatively high speed (e.g., 1–2 meters per second) which increases the rate of sweat evaporation from the skin and minimizes occupant discomfort when their skin perspires. Air movement not only increases the evaporative rate at the skin surface (in higher temperatures), it also determines the convective heat and mass exchange of the human body with the surrounding air, which affects thermal comfort.

Night-time ventilation is a strategy used to cool down the building structure via lower external air temperatures at night, providing the building has a high thermal mass and exposed structure. This strategy employs the building's thermal mass as an intermediate storage medium, allowing the structure to absorb the heat built up during the day and to flush it away during the night. In the appropriate climate this is a suitable solution for office buildings which are typically occupied during the daytime only. As the coolness is stored in the thermal mass (e.g., exposed concrete), it reduces peak daytime temperatures, reduces/eliminates dependence on artificial cooling, enhances internal comfort conditions during occupancy periods, and moderates temperature fluctuations inside the building. Another advantage of night-time ventilation is the purge and decrease of indoor pollutants.

**Night-time ventilation is a strategy used to cool down the building structure via lower external air temperatures at night, providing the building has a high thermal mass and exposed structure.**

---

[1] It should be noted, as mentioned previously, that "hybrid" buildings will still require a mechanical system and plant equipment. The embodied energy required for the mechanical plant equipment and the loss of floor space can have a significant impact on the building.

# 2.0 Case Studies

# Project Data:

**Year of Completion**
- ▸ 1996

**Height**
- ▸ 127 meters

**Stories**
- ▸ 31

**Gross Area of Tower**
- ▸ 36,000 square meters

**Building Function**
- ▸ Office

**Structural Material**
- ▸ Concrete

**Plan Depth**
- ▸ 8 meters (from central core)

**Location of Plant Floors:**
- ▸ 19

# Ventilation Overview:

**Ventilation Type**
- ▸ Mixed-Mode:
  Complementary-Changeover

**Natural Ventilation Strategy**
- ▸ Wind-Driven Single-Sided Ventilation

**Design Strategies**
- ▸ Double-skin façades
- ▸ "Fish Mouth" device which adjusts air intake speed
- ▸ Aerodynamic external form

**Double-Skin Façade Cavity:**
- ▸ Depth: 500 mm
- ▸ Horizontal Continuity: 2 meters
- ▸ Vertical Continuity: 3.5 meters (floor-to-floor)

**Approximate Percentage of Year Natural Ventilation can be Utilized:**
- ▸ 75%

**Percentage of Annual Energy Savings for Heating and Cooling:**
- ▸ Unpublished

**Typical Annual Energy Consumption (Heating/Cooling):**
- ▸ Unpublished

Case Study 2.1

# RWE Headquarters Tower Essen, Germany

## Climate

Located on the River Ruhr in the North Rhine Westphalia area of Germany, the climate of Essen is influenced by its proximity to the northern coastline. One of the warmest cities in Germany, Essen experiences its mild weather largely due to the westerly sea breezes that blow in from the North Sea. Winter is fairly mild, experiencing an average daytime temperature of 5 °C with only the occasional frost and snow. The warmer summer months average 22 °C and can bring unpredictable rain showers, although short lived (see Figure 2.1.1).

## Background

RWE Headquarters Tower is conceptually composed of a base, a shaft, and a capital (see Figure 2.1.2). The cone-shaped base is only visible from the garden as a rising terrain completely conceals it at the entry. Facilities that could not have been adequately accommodated in the slender tower (the casino, restaurant, and conference rooms) were placed in an adjacent seven-story building acting as a gateway to the tower's main entrance.

The tower itself has a circular plan with a service and circulation core at its center, and offices spaces located along its perimeter (see plan and section, Figures 2.1.3 & 2.1.4). Elevators are mostly detached from the office space in a separate form attached to the southeast side of the tower, though there is vertical circulation within the circular plan at the north and south of the tower. This approach allows the central space to be open and free of obstructions. With the office spaces located along the building perimeter, access to daylight and natural ventilation are maximized. A double-height floor housing the mechanical equipment is located on level 19.

## Natural Ventilation Strategy

The tower features a double-skin façade consisting of extra-clear toughened exterior glass panels (measuring approximately 2 meters wide by 3.5

## Climatic Data:[1]

**Location**
- ▸ Essen, Germany

**Geographic Position**
- ▸ Latitude 51° 27′ N, Longitude 7° 0′ E

**Climate Classification**
- ▸ Temperate

**Prevailing Wind Direction**
- ▸ West-southwest

**Average Wind Speed**
- ▸ 2.7 meters per second

**Mean Annual Temperature**
- ▸ 10 °C

**Average Daytime Temperature during the Hottest Months (June, July, August)**
- ▸ 22 °C

**Average Daytime Temperature during the Coldest Months (December, January, February)**
- ▸ 5 °C

**Day/Night Temperature Difference During the Hottest Months**
- ▸ 9 °C

**Mean Annual Precipitation**
- ▸ 934 millimeters

**Average Relative Humidity**
- ▸ 73% (hottest months); 82% (coldest months)

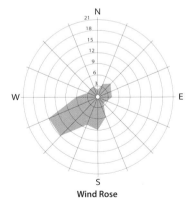

Wind Rose

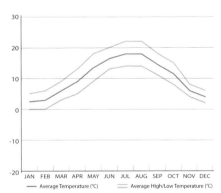

Average Annual Temperature Profile (°C)

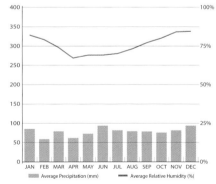

Average Relative Humidity (%) and Average Annual Rainfall

▲ Figure 2.1.1: Climate profiles for Essen, Germany.[1]
◀ Figure 2.1.2: Overall view. © Christian Voermann

[1] The climatic data listed for Essen was derived from the World Meteorological Organization (WMO) and Deutscher Wetterdienst (German Weather Service).

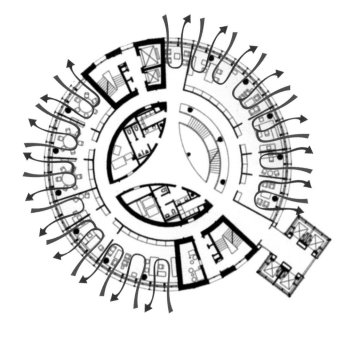

### Plan

The natural ventilation strategy relies on single-sided ventilation in each office through a double-skin façade that controls high direct wind speeds. The cellular offices are naturally ventilated through floor-to-ceiling sliding glass panels on the inner face of the double-skin façade. Every office space has at least one operable window.

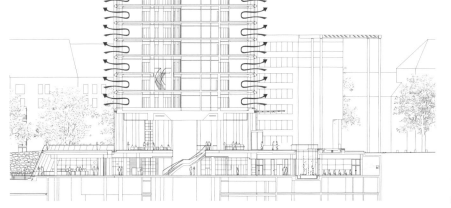

### Section

Cellular offices and meeting rooms are placed at the building perimeter. Single-sided natural ventilation enters the space through sliding glass panels in the inner face of the double-skin façade.

▲ Figure 2.1.3: Typical office floor plan.
◀ Figure 2.1.4: Building section. (Base drawings © ingenhoven architects)

▲ Figure 2.1.5: Exterior view of the double-skin façade. Note the air exhaust grills in the underside of the Fish Mouth devices in the cavity at every other module. Also note the vertical glass fins segmenting the cavity.
© ingenhoven architects

high) (see Figure 2.1.5), a 500 mm air cavity, and an inner façade featuring floor-to-ceiling sash windows with fixed and sliding glass panels that can be opened up to approximately 150 mm with a winding handle. For acoustic insulation and ventilation purposes, the intermediate air cavity is segmented vertically at each floor by a "Fish Mouth" device and horizontally with a glass panel (see Figure 2.1.6). The Fish Mouth device was specially designed for fresh air intake and exhaust, control of vertical sound transmission, and support of the exterior skin of the façade. Together with the double-skin façade, the Fish Mouth comprises a key element of the tower's natural ventilation strategy.

There are actually two types of the Fish Mouth device; one type emits air into the façade cavity while the other exhausts stale air (see Figure 2.1.7). Each double-skin module has only one type of the Fish Mouth device providing a module that emits fresh air and a

module that exhausts stale air. Each façade module alternates along the entire envelope, allowing even the smallest office with only two façade modules to have access to both devices. This alternating layout, separated by vertical glass fins, prevents stale air from re-entering the adjacent façade cavity and short-circuiting problems where fresh air would exit the cavity without ventilating the office spaces.

Another feature of the Fish Mouth is to adjust the air speed as it passes through the device, making it slower or faster, depending on the exterior wind speed and desired ventilation rates. Since wind speeds increase at higher altitudes, the Fish Mouth device moderates wind pressures across the envelope through two opening sizes (one size for above floor 16 and one for below). CFD simulations and large-scale wind tunnel tests were used to optimize the design, form, and opening size of the Fish Mouth device. The tapered shape

**The intermediate air cavity is segmented vertically at each floor by a Fish Mouth device specially designed for fresh air intake and exhaust, control of vertical sound transmission, and support of the exterior skin of the façade.**

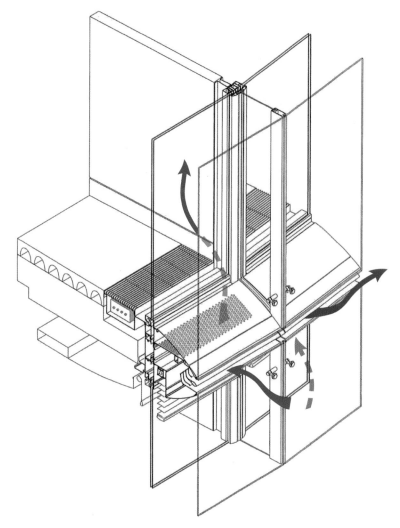

▲ Figure 2.1.6: An isometric drawing of the double-skin façade showing the Fish Mouth inlet device.
© ingenhoven architects

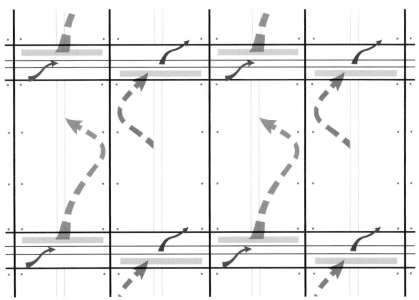

▲ Figure 2.1.7: Façade elevation showing the diagonal flow of air across bays. © ingenhoven architects

of the device with curved reflective surfaces also maximizes the admittance of natural light by approximately 20 percent (Evans 1997).

## Mixed-Mode Strategy

The building has a Complementary-Changeover system that switches between mechanical and natural ventilation on a seasonal or daily basis as required. A custom solution was developed which enables the building to utilize natural ventilation for 70–80 percent of the year (Briegleb 2000). Instead of a conventional mechanical system used for heating and cooling, a convector system supplements natural ventilation. The convector system consists of water pipes located at the edge of the floor plate and integrated into ceiling panels.

During the cooling season, cold water at a maximum output of 16–18 °C flows through these pipes. In an office where a window has been opened, the building management system (BMS) automatically stops the cooling ceiling elements to combat condensation. When the windows cannot be opened for natural ventilation, the building is sealed and mechanical ventilation is activated. In this scenario, air is supplied (at a minimum ventilation rate of 2.5 air changes per hour) and returned from ducts in the circulation corridor at the center of the building (Evans 1997).

## Interface with the Central Building Management System

The building's environment is consistently regulated through the central BMS, which receives information from sensors located in each room and on the exterior of the building. The BMS adjusts the blinds in the cavity, controls ventilation openings, and activates

▲ Figure 2.1.8: Control panel next to office doors. © ingenhoven architects

▲ Figure 2.1.9: Interior view showing perforated aluminum blinds in the façade cavity. © ingenhoven architects

the mechanical ventilation system if required. In extreme conditions (such as wind speeds of 7 m/s below floor 16 and 10 m/s above), the BMS automatically closes ventilation openings on the outer skin and activates the mechanical ventilation system. Conversely, when a window is opened in favorable weather conditions, the BMS automatically turns off the convector in the specific room with the open window.

Occupants of RWE Headquarters Tower have a high level of control over their environment. Each office has a minimum of two sliding windows that can be directly controlled by the occupants. A control panel next to each office door (see Figure 2.1.8) allows the individual to override the BMS by adjusting the temperature in their immediate area (up to a three-degree variation from the rest of the building) and the illumination level through automatically controlled blinds in the façade cavity (see Figure 2.1.9).

## Other Sustainable Design Elements

The double-skin façade also acts as a thermal and acoustic buffer protecting against heat gain/loss and high wind speeds. Perforated aluminum blinds located in the façade cavity protect against solar heat gain and glare. Additional shading and protection from solar heat gain is provided by translucent roller screens installed on the interior side of the inner skin. This efficient use of daylight reduces dependence on artificial lighting and minimizes the cooling loads due to internal heat gains.

The building's thermal performance is influenced by partially exposed concrete ceiling slabs which absorb excess heat during the day and emit the stored heat at night through night-time flush ventilation in summer. This moderates internal temperature fluctuations and reduces mechanical heating/cooling loads during peak hours. By activating

**As wind speeds increase at higher altitudes, the Fish Mouth device moderates wind pressures across the envelope through two opening sizes (one size for above floor 16 and one for below).**

**The segmentation of the façade cavity may restrict the free flow of air around the building. Without this segmentation, the double-skin façade could reduce and balance pressure differentials across openings.**

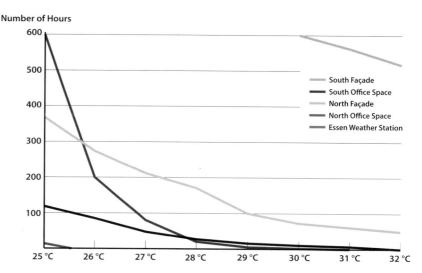

▲ Figure 2.1.10: Results of a study which measured the internal temperature of the façade cavity against the interior room temperatures which resulted in minimal impact. (Source: Pasquay 2004)

the building's thermal storage capacity along with night-time cooling the energy requirements were 25 percent lower than those of a conventionally air-conditioned building (Briegleb 2000). Perforated ceiling panels allow for some exploitation of the slab's thermal storage capacity. Water pipes for radiant cooling/heating are also incorporated in the ceiling.

### Analysis – Performance Data

Post-occupancy studies were carried out over 1998–2000, including recording room temperatures in relation to exterior and façade temperatures as well as air change rates. One such study (see Figure 2.1.10) showed typical room temperatures did not exceed 28 °C for a period in May in natural ventilation mode without the use of the cooled ceiling (Pasquay 2004). It is important to calculate the exterior surface temperature of the façade and the internal temperature of the façade cavity, as these can marginally affect interior room temperatures. Even though the façade experienced a

higher percentage of excessive heat during this study, the interior office spaces remained quite comfortable.

Another post-occupancy study measured the natural air change rate per hour for a typical office on the first floor and the sixteenth floor (see Figure 2.1.11). The resulting tracer gas measurement of the natural air change rate per hour averaged a little over 6 ACH, which is well above the designed minimum of 2.5 ACH.

### Analysis – Strengths

▸ Compared to other buildings, a circular building form has the largest ratio of floor area to perimeter length. This reduced surface area of the envelope aids in a reduced heat gain/loss that occurs through the external skin.

▸ The aerodynamic cylindrical form of the tower encourages wind flow around the façade and minimizes wind loads on the envelope, thus

facilitating natural ventilation within the building.

▸ The double-skin façade moderates wind pressure and velocity on the internal windows, enhancing the prospects for natural ventilation, particularly at the top floors where external wind speed is high.

▸ In response to increasing wind speeds at higher altitudes, the intake and exhaust openings of the Fish Mouth devices within the double-skin façade change in size according to the various pressure coefficients occurring on the building envelope.

▸ The Fish Mouth device prevents rain from entering the building, improves daylighting in the offices, and restricts vertical sound transmission through the façade cavity.

▸ The aerodynamic form of the Fish Mouth device regulates airflow, which reduces the risks associated with natural ventilation in tall buildings due to variations in wind speeds. This also minimizes control difficulties caused by the wide range of pressure differentials that occur across envelope openings in tall buildings.

## Analysis – Considerations

The following areas could be causes for concern if adopting similar strategies in other buildings and should therefore be considered:

- ▶ The segmentation of the façade cavity may restrict the free flow of air around the building. This may result in a wide range of airflow rates that are difficult to control, resulting in some rooms being better ventilated than others. Without this segmentation, the double-skin façade could reduce and balance pressure differentials across openings and bring the pressure gradient to a state of equilibrium.

- ▶ The circular form of the building causes the Fish Mouth openings pointing directly into the wind to have the highest natural ventilation air change rates, whereas Fish Mouth openings that are located at a 90° angle to the wind direction may see minimal air changes.

- ▶ Short-circuiting of the ventilation route may occur, especially closer to the façade zone, leading to insufficient ventilation of the inner zone of the room.

- ▶ The interior may over-cool during certain periods of the year since night ventilation is not automatically controlled by the central BMS.

- ▶ Office layout flexibility may be restricted due to the façade cavity

module, which is segmented at interior partitions and requires two or more Fish Mouth devices (one inlet and one outlet) for every room.

## Project Team

**Owner/Developer:** Hochtief AG
**Design Architect:** ingenhoven architects
**Structural Engineer:** Buro Happold; Hochtief AG
**MEP Engineer:** HL-Technik AG Beratende Ingenieure; IGK Ingenieurgemeinschaft Kruck; Buro Happold
**Environmental Engineer:** Geocontrol Umwelttechnische Beratung
**Main Contractor:** Hochtief AG
**Other Consultants:** Josef Gartner and Company (Façade Consultant)

**References & Further Reading**

**Books:**

- ▶ Briegleb, T. (ed.) (2000) *Ingenhoven Overdiek and Partners: High-Rise RWE AG Essen.* Birkhäuser: Basel.

- ▶ Feireiss, K. (2002) *Ingenhoven Overdiek and Partners, Energies.* Birkhäuser: Basel, pp. 214–241.

- ▶ Ingenhoven, C. (2001) "Greening office towers," in Beedle, L. (ed.) *Cities in the Third Millennium – Proceedings of the CTBUH 6th World Congress.* Spon Press: London, pp. 527–530.

**Journal Articles:**

- ▶ Evans, B. (1997) "Through the glass cylinder," *Architect's Journal,* vol. 205, no. 19, pp. 42–45.

- ▶ Pasquay, T. (2004) "Natural ventilation in high-rise buildings with double façades, saving or waste of energy," *Energy and Buildings,* vol. 36, no. 4, pp. 381–389.

- ▶ Pearson, J. (1997) "Delicate Essen," *Architectural Review,* vol. 202, no. 1205, pp. 40–45.

- ▶ Pepchinski, M. (1997) "RWE AG Hochhaus – Essen, Germany," *Architectural Record,* vol. 185, no. 6, pp. 144–151.

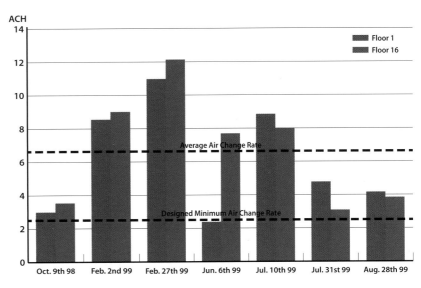

▲ Figure 2.1.11: Results of a study which measured the natural air change per hour as tested in a typical office on the first and sixteenth floors. (Source: Pasquay 2004)

## Project Data:

**Year of Completion**
- ▸ 1997

**Height**
- ▸ 259 meters

**Stories**
- ▸ 56

**Gross Area of Tower**
- ▸ 85,500 square meters

**Building Function**
- ▸ Office

**Structural Material**
- ▸ Composite

**Plan Depth**
- ▸ 16.5 meters (from central void)

**Location of Plant Floors:**
- ▸ B2–G, 4–5, 51–56

## Ventilation Overview:

**Ventilation Type**
- ▸ Mixed-Mode: Complementary-Changeover

**Natural Ventilation Strategy**
- ▸ Cross and Stack Ventilation (connected internal spaces)

**Design Strategies**
- ▸ Double-skin façades
- ▸ Stepping sky gardens connected by segmented central atrium
- ▸ Small aerofoil sections above/below ventilation slots in façade

**Double-Skin Façade Cavity:**
- ▸ Depth: 200 mm
- ▸ Horizontal Continuity: 1.5 meters
- ▸ Vertical Continuity: 2.4 meters (between floor spandrel panels)

**Approximate Percentage of Year Natural Ventilation can be Utilized:**
- ▸ 80%

**Percentage of Annual Energy Savings for Heating and Cooling:**
- ▸ 63% compared to a fully air-conditioned German office building (measured)
- ▸ 38% compared to a fully air-conditioned office building, built to EnEV 2007, the 2007 German Energy Conservation Ordinance (measured)

**Typical Annual Energy Consumption (Heating/Cooling):**
- ▸ 117 kWh/m$^2$ (measured)

# Commerzbank Frankfurt, Germany

## Local Climate

Frankfurt is located in a warm temperate climate which experiences relatively mild weather all year round. Although fairly uncommon, short-lived extremes occasionally occur. Summers are typically warm and sunny with a chance of light rain occurring. On the sunniest of days, the temperature often rises above 24 °C, even topping 30 °C on occasion, and drops to around 15 °C in the evenings. During the winter season, January is characteristically the coldest month. Daytime temperatures hover around 4 °C and at night drop just below freezing, when a light snow is likely (see Figure 2.2.1).

## Background

Commerzbank was completed in 1997 and is still considered one of the most ecological tall buildings ever built. The building's green credentials were a socially responsible aspiration of the owner-occupier and an essential pre-condition for granting the building's generous 14:1 plot ratio by city

authorities (typically the building would have been governed by a 5:1 plot ratio). The tower is centrally located within the densely built-up financial district of Frankfurt am Main (see Figure 2.2.2). At the pedestrian scale, the tower is set back from a six-story-high base which includes a plaza, a bank, a 500-person auditorium, a multi-story garage, shops, and 27 residences. The use of this residential perimeter block allows the tower to integrate with adjacent low-rise buildings, some of which date back to the early 1870s.

The building has a triangular plan with a central atrium running the full height of the building but divided into four segments, around which sky gardens and office spaces are arranged in a spiraling configuration (see plan and section, Figures 2.2.3 & 2.2.4). The three corners of the triangle house the main structural elements and include vertical circulation, toilets, and other service facilities. The structural system consists of three atrium columns with peripheral girders and six mega columns at the corners of the building, connected by eight-story vierendeel girders.

## Climatic Data:[1]

**Location**
- ▸ Frankfurt, Germany

**Geographic Position**
- ▸ Latitude 50° 7′ N, Longitude 8° 41′ E

**Climate Classification**
- ▸ Temperate

**Prevailing Wind Direction**
- ▸ South-southwest

**Average Wind Speed**
- ▸ 4 meters per second

**Mean Annual Temperature**
- ▸ 10 °C

**Average Daytime Temperature during the Hottest Months (June, July, August)**
- ▸ 24 °C

**Average Daytime Temperature during the Coldest Months (December, January, February)**
- ▸ 4 °C

**Day/Night Temperature Difference During the Hottest Months**
- ▸ 11 °C

**Mean Annual Precipitation**
- ▸ 621 millimeters

**Average Relative Humidity**
- ▸ 53% (hottest months); 76% (coldest months)

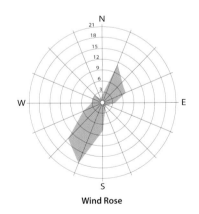

Wind Rose

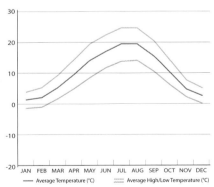

Average Annual Temperature Profile (°C)

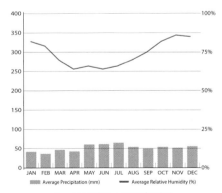

Average Relative Humidity (%) and Average Annual Rainfall

▲ Figure 2.2.1: Climate profiles for Frankfurt, Germany.[1]
◄ Figure 2.2.2: Overall view from the south. © Marshall Gerometta/CTBUH

[1] The climatic data listed for Frankfurt was derived from the World Meteorological Organization (WMO) and Deutscher Wetterdienst (German Weather Service).

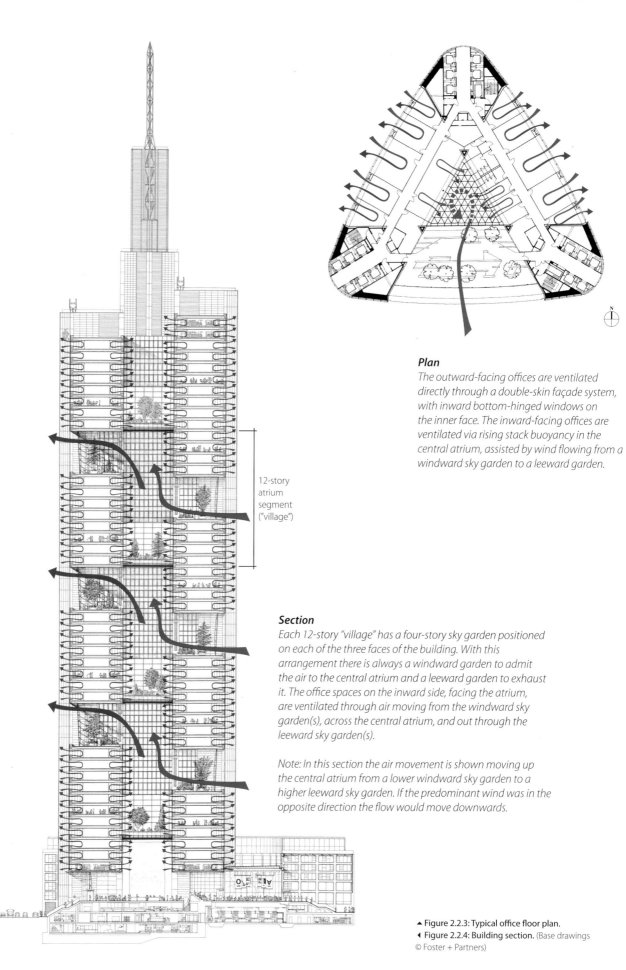

### Plan

The outward-facing offices are ventilated directly through a double-skin façade system, with inward bottom-hinged windows on the inner face. The inward-facing offices are ventilated via rising stack buoyancy in the central atrium, assisted by wind flowing from a windward sky garden to a leeward garden.

12-story atrium segment ("village")

### Section

Each 12-story "village" has a four-story sky garden positioned on each of the three faces of the building. With this arrangement there is always a windward garden to admit the air to the central atrium and a leeward garden to exhaust it. The office spaces on the inward side, facing the atrium, are ventilated through air moving from the windward sky garden(s), across the central atrium, and out through the leeward sky garden(s).

Note: In this section the air movement is shown moving up the central atrium from a lower windward sky garden to a higher leeward sky garden. If the predominant wind was in the opposite direction the flow would move downwards.

▲ Figure 2.2.3: Typical office floor plan.
◀ Figure 2.2.4: Building section. (Base drawings © Foster + Partners)

▲ Figure 2.2.5: View up the central atrium showing the steel and glass diaphragms which segment the building into 12-story "villages." These diaphragms act as a boundary to limit stack pressures and smoke spread.
© Nigel Young / Foster + Partners

**Each 12-story village has three sky gardens positioned on each of the three faces of the building so there is always a windward garden to admit air to the central atrium and a leeward garden to exhaust it.**

## Natural Ventilation Strategy

The building is divided into 12-story "villages" stacked on top of each other,[2] with a central atrium segmented through the use of steel and glass diaphragms at the boundary level of each village (see Figure 2.2.5). These diaphragms limit stack pressures and smoke spread in the central atrium, isolating each village to be ventilated completely independently from the others. Each village has a four-story sky garden positioned on each of the three faces of the building, configured in a spiraling pattern up the building, thus allowing ventilation to occur regardless of wind direction as there is always a windward garden to admit the air to the central atrium and a leeward garden to exhaust it (depending on the wind direction, sometimes the air travels up the village's atrium, and sometimes down). The atrium and sky gardens can be considered as sheltered,

quasi-external spaces that moderate between external and internal environments. Part of the strategy for these spaces – as well as assisting in the natural ventilation strategy – is that they serve as important social/amenity spaces for the building occupants (see Figure 2.2.6).

Although a large part of the strategy for this design was inspired by the socially responsible approach of the building client-developer, it should be noted that much of it was due to German planning regulations, which specified that all permanent workplaces have direct access to view and light by being positioned not farther than seven meters from the façade. This is accomplished through the 16.5-meter-wide office "wings" which span between two corners of the triangular plan. The wings are split by a wide central corridor with half the offices facing the central atrium (see Figure 2.2.7), and the other half

facing the exterior. This solution permits each cellular office, regardless of inward or outward orientation, to have access to outside air, light, and views.

The external-facing offices are ventilated directly from the external envelope through a double-skin façade system. The outer layer of this is composed of a solid pane of laminated glass which deflects strong wind and rain. The 200 mm façade cavity spans between the floor spandrel panels (and is thus not continuous beyond one floor in the vertical direction or beyond one façade module in the horizontal direction). The cavity is ventilated at the top and bottom through 125 mm continuous slots in the external façade. Small aerofoil-section strips are positioned at sill levels just above/below these ventilation slots to improve airflow through the cavity (capturing air flowing up or down the face of the building and directing it into the cavity)

---

[2] There are three typical 12-story "villages" above the main lobby area (starting at the seventh floor). A fourth "village" of two, rather than three wings, ten floors in height, including two mechanical floors, extends beyond the highest typical village (see Figure 2.2.2). This topmost village is thus an open "V" sat on the triangular plan below and fully open to the outside. The west corner core also extrudes higher still into a pinnacle housing additional mechanical equipment and antenna.

▲ Figure 2.2.6: View of a high-level sky garden which also serves as social/amenity space in the building.
© Nigel Young / Foster + Partners

▲ Figure 2.2.7: View from an inward-facing office. © Nigel Young / Foster + Partners

▲ Figure 2.2.8: Exterior façade detail showing the 125 mm ventilation slots with aerofoil-strips protruding above/below them to facilitate airflow through the double-skin cavity. © Ralph Richter/archenova

and to avoid the short-circuiting of air (see Figures 2.2.8 & 2.2.9). The inner face of the double-skin façade consists of bottom-hinged, double-glazed windows that open inwards at the top to a maximum 15° angle (see Figure 2.2.10), thus allowing air from the cavity to flow directly into the space. The double-skin façade acts mostly as a buffer to control wind-driven air into the office space, and partly as a thermal flue by stack buoyancy of rising warmer air in the façade cavity drawing out stale air from the offices. However, the limited height of the cavity limits the effectiveness of this latter strategy.

The office spaces on the inward side, facing the atrium, are ventilated (via bottom-hinged windows facing the atrium) via air moving through the sky gardens as previously described. Each 14-meter-high sky garden façade has large motorized pivoting windows at the bottom for air intake and top for air extraction (see Figure 2.2.11). Primarily through negative pressures on the leeward face of the building and rising air in the central atrium through the stack effect, the air is pulled across the building plan and section from the windward side, drawing air from the offices on the inward face in turn.

The ventilation strategies for the inward-facing offices, outward-facing offices, and central corridors work independently of each other. As a result, the inward-facing offices (which have shown to be most successful) are able to be naturally ventilated year-round, regardless of the ventilation strategy employed by the outward-facing offices.

Overall, the natural ventilation strategy of the offices can be characterized as single-sided ventilation. The natural ventilation strategy for the atria can be characterized as a combination of cross and stack ventilation, which are key for providing proper natural ventilation conditions for the internal offices.

## Mixed-Mode Strategy

Comfort was a critical factor at the design stage and to solely rely on natural ventilation all year round to maintain ideal comfort was considered difficult – especially in Frankfurt's occasional high summer temperatures. The building was therefore designed with a Complementary-Changeover system which switches between mechanical and natural ventilation on a seasonal (or even daily) basis.

Overall, the building was designed to be naturally ventilated for approximately 60 percent of the year (although actual performance has exceeded this goal, see "Analysis – Performance"). Mechanically assisted air-conditioning was intended to only be activated under extreme conditions (i.e., when the wind is too strong or when external temperatures are either too high or too low to allow windows to be opened). Mechanical ventilation introduces conditioned air into the offices via a system of supply ducts located above the office ceiling panels. The intake air is cooled in the summer and heated in the winter to maintain thermal comfort. Exhaust air is drawn back to the main plant at the mechanical floors (located at the top of the building) through the plenum created by the suspended ceiling.

On hot summer days a water-filled cooling system integrated in the ceiling panels of the office spaces provides additional cooling (primarily to offset the internal heat gains from occupants, as the air-conditioning system only provides air cooling to the design temperature of 26 °C, but not lower). In winter, supplemental heating is provided through panel

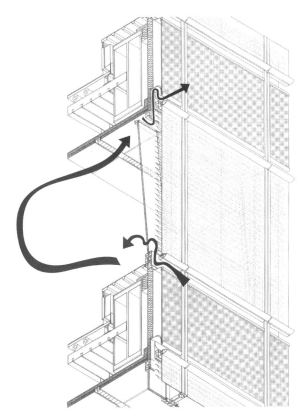

▲ Figure 2.2.9: Isometric drawing of a typical bay in the external office cladding showing airflow through the double-skin façade, entering the space around the sides of the open window and exiting around the sides at the top. © Foster + Partners

radiators located beneath perimeter windows in office areas (see Figure 2.2.12) and under-floor heating in the sky gardens. Sky gardens always remain naturally ventilated and intermediate in temperature between the interior and the exterior, thus functioning as semi-outdoor spaces. In contrast, the central corridor zones between office spaces in the office wings are mechanically ventilated at all times through an air supply and exhaust system.

## Interface with the Central Building Management System

The central BMS controls the operation of motorized windows, blinds, chilled ceilings, air-conditioning, and perimeter heating according to an "intelligent algorithm" that switches modes for an optimum balance between occupant comfort and energy efficiency. In addition, it can completely deactivate the air-conditioning for areas of the building not being used. The system is zoned according to the 12-story "village"

**The outward-facing offices benefit from direct access to a simple double-skin façade natural ventilation system with inward-pivoting windows which allows single-sided ventilation through wind/pressure-driven ventilation.**

▲ Figure 2.2.10: The inner face of the double-skin façade consists of bottom-hinged, double-glazed windows that open inwards at the top. © Ralph Richter/archenova

▲ Figure 2.2.11: Each sky garden façade has large motorized pivoting windows in the lower segment of the façade for air intake and at the top for air extraction. Note the fully open windows at the top of the façade in this image. © Ralph Richter/archenova

horizons and was originally informed by nine weather sensors in each of the sky gardens (though this has since been changed, see "Post-Completion Changes/Improvements"). The BMS thus controls the internal climate system according to the number of people in the building, the usage of the system, preset interior temperature limits (26 °C maximum in summer and 17 °C minimum in winter for offices) as well as wind, solar intensity, and humidity measurements from the weather sensors throughout the building. It should be noted that the sky gardens are not controlled, though temperatures rarely go above 27 °C in summer. The BMS also determines the level of control occupants have over the building according to external weather conditions. Occupants have control over their environment and can override BMS-operated systems such as lighting and blinds through wall-mounted switches located next to each office

door (see Figure 2.2.13). Furthermore, an indicator light tells occupants when the system is operating in mechanical or natural ventilation mode. When in natural ventilation mode, occupants have further control and are free to open a window. The windows only shut automatically when external wind speeds are too high (15 m/s only for windows facing the prevailing wind), or when external temperatures are above 26 °C or internal temperatures below 18 °C.

## Other Sustainable Design Elements

In addition to the natural ventilation strategy, other sustainable features were implemented in the design. During summer when external temperatures do not allow for the use of natural ventilation, a water-based chilled ceiling panel system provides cooling in the office areas. All work spaces utilize

satisfactory levels of daylight. Direct sunlight is blocked through the use of motorized blinds integrated in the intermediate space of the double-skin façade. The building also makes extensive use of efficient artificial lighting systems, including the use of light sensors. The gray water from the cooling towers is used to flush toilets.

## Analysis – Performance

Energy consumption analysis from 1999–2008 was compared to some of the toughest national energy benchmarks to date: "EnEV 2007" – a German Energy Conservation Ordinance. This comparison shows the building surpassing those benchmarks even though they were written ten years after Commerzbank was built (see Figure 2.2.14). While varying slightly over the last ten years, the total annual thermal energy consumption ranges between 105 and 128 kWh/m². It is important to mention that these figures fall below all EnEV 2007 benchmarks including fully air-conditioned offices (190 kWh/m²), offices mechanically ventilated and heated but not cooled (160 kWh/m²) and most surprisingly naturally ventilated office buildings (135 kWh/m²) (Gonçalves 2010). While natural ventilation is attainable for as much as 80 percent of the year (+/- 5 percent), some zones in the building benefit from natural ventilation during the entire period of occupation annually (Gonçalves 2010).

Compared to a standard building on the Commerzbank site, the sustainable features also show a significant increase on the investment return for the project (see Figure 2.2.15).

It is also worth noting that the building was one of the first major commercial buildings to employ computational fluid dynamic (CFD) analysis. Simulations were carried out at the design stage to model airflow patterns around and through the building, as well as to predict the velocity, direction, and temperature of the air in the atrium and sky gardens at different times of the year and under a variety of weather conditions. In addition, CFD was used to test the effect of air on the temperature of adjacent spaces and surfaces. The results of these simulations were then used to inform the design with respect to the size of the external ventilation openings for the purposes of optimizing airflow through the building.

### Post-Completion Changes/ Improvements

Since 2002, the satisfactory conditions of the microclimate inside the villages meant that the control of the windows of the internal-facing offices were left to the occupants rather than the BMS,

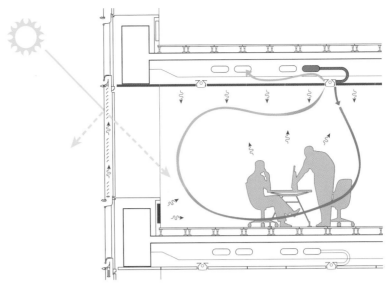

▲ Figure 2.2.12: Diagram showing a typical office during either hot or cold weather extremes that prohibit natural ventilation. Mechanical ventilation introduces air which on hot summer days is chilled by a water-filled cooling system integrated in the ceiling panels. In winter, supplemental heating is provided via panel radiators located beneath perimeter windows. © Foster + Partners

▲ Figure 2.2.13: Wall-mounted switches next to each office door. © Ralph Richter/archenova

as the occupants were used to opening the windows for natural ventilation. Interestingly, with more control over their naturally ventilated environment, some occupants grew more tolerant to internal summer temperatures in excess of the 26 °C design temperature – and preferred this rather than closing the windows and switching to artificial cooling. The control of the internal environmental conditions was thus simplified as the building became increasingly "manual," resulting in higher energy savings (Gonçalves & Bode 2010).

Since 2008, there has been an increase in the number of work stations from 2,400 to 2,820. This increase in population led to changes in the internal layout of the usable floors. The external-facing offices were combined with the central corridor to create more of an open-plan layout in which that entire space is now naturally ventilated using the same single-sided system through the double-skin façade, and has shown to continue to work effectively even with this increased depth.

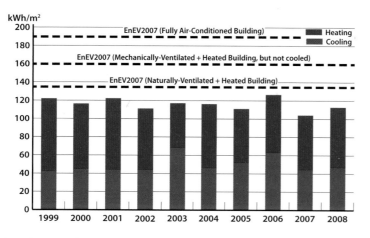

▲ Figure 2.2.14: Average yearly energy consumption for heating and cooling Commerzbank, compared to EnEV 2007 German energy conservation benchmarks. (Source: Gonçalves 2010)

| | Typical building on Commerzbank site | Commerzbank Headquarters Building |
|---|---|---|
| Net Area | 21,800 m² | 63,500 m²* |
| Building Cost | £67M | £203M |
| Investment Value | £100M | £286M |
| Investment Return | £10M | £83M |

*Net Area for these figures considers only the usable area within the tower footprint (i.e., offices, meeting rooms, internal circulation areas, etc. and excludes lift lobbies, staircases, toilets, entrance hall, gardens, etc.).

▲ Figure 2.2.15: Approximation of the investment return on the sustainable features in comparison to a standard building at the Commerzbank site (These figures take into account the higher permissible plot ratio through adopting sustainable features). (Source: Grontmij – formerly Roger Preston & Partners)

**The amount of leasable floor area occupied by the atria and sky gardens is considerable and may not be viable on a more commercial, or speculative, office building.**

After a decade of monitoring, the building's energy consumption shows how natural ventilation in a tall office building can substantially minimize energy use and offset changes that might incur an increase in energy use such as the increase in occupancy density. Also, in 2002, the nine weather sensors that were originally located in each sky garden were replaced by a single one at the top of the building. This was done to achieve better accuracy in recording wind data compared with the measurements taken from inside the gardens and, therefore, achieved more precision in the control of the windows on the three external faces (Gonçalves & Bode 2010).

### Analysis – Strengths

▸ Natural ventilation is facilitated by the building layout which moves away from the conventional central core and surrounding office

arrangement. Pushing typical core functions to the outer corners of the building gives more flexibility in space planning, allowing for the creation of a central atrium and spiraling sky gardens that permit the entry of fresh air (and natural light) into the inward-facing offices.

▸ The outward-facing offices benefit from direct access to a double-skin façade natural ventilation system. The inward-pivoting windows of the double-skin allow for single-sided, pressure-driven ventilation which originates from the façade cavity and enters the office spaces at acceptable velocities.

▸ Each of the 12-story villages contains three sky gardens, one on each face of the building. In this manner, the spatial arrangement of the spiraling sky gardens allows for efficient cross-ventilation through the central part of the building as there is always a windward garden to admit fresh air and a leeward garden to exhaust it.

▸ As each atrium creates stack effect and supplements wind-driven cross-ventilation through the sky gardens, the system is still effective with little or no wind. Natural ventilation on the inward-facing offices is so successful they can be naturally ventilated year-round, even when the external-facing offices have their windows closed.

▸ The sky gardens act as solar collectors and thermal buffers between the interior and exterior, preheating the air before it enters the office spaces in the winter. Also, the sky gardens are an important element of the ventilation strategy, acting as extraction chimneys which use stack effect to exhaust air from the building. In addition, they provide important communal spaces that foster social interaction and act as destination and transition spaces that ease occupant circulation throughout the building. In this sense, the sky gardens function as central orientating spaces which offer visual connectivity and tremendous social and psychological benefits.

▶ The vertical segmentation of the atrium (into 12-story-high segments) through the use of glass and steel diaphragms prevents the development of extreme stack flows and drafts due to large pressure differentials between the top and bottom of the atrium. The segmentation of the atrium also provides fire separation and ventilation zones.

## Analysis – Considerations

The following areas could be causes for concern if adopting similar strategies in other buildings and should therefore be considered:

▶ The configuration and layout of the building does not specifically acknowledge prevailing winds or the asymmetry of the sun's path. Without a proper control strategy implemented by the BMS, strong winds blowing from a specific direction could cause the outward office wings facing the prevailing wind to be better ventilated than the outward offices on the leeward face. The layout of the plan could lead to uneven pressure distribution and ventilation rates in various wings of the building.

▶ The bottom-hung windows in the outward-facing offices may not allow sufficient air to enter the office from the double-skin façade

cavity, with airflow remaining predominantly in the façade zone.

▶ The amount of floor space occupied by atria and sky gardens is considerable and may not be viable on a more commercial, or speculative, office building. The design also requires a significantly larger façade area for the total floor area (due to the inner atrium-facing façades) as compared to a typical office building, further increasing costs.

▶ With sky garden windows closed during winter, there may not be sufficient cross flow to induce substantial ventilation of inward offices. However, these spaces have lower fresh air requirements in winter, thus with the volume of the gardens and atrium, combined with natural infiltration, sufficient air changes have shown to be provided.

▶ Though the atrium is compartmentalized into 12-story sections for smoke spread, there is still the potential for significant smoke/fire spread within the villages due to the amount of open atrium space.

## Project Team

**Owner/Developer:** Commerzbank
**Design Architect:** Foster + Partners
**Structural Engineer:** Arup; Krebs und Kiefer
**MEP Engineer:** Grontmij (formerly Roger Preston & Partners); Pettersson & Ahrens; Schad and Holzel
**Project Manager:** Nervus GmbH
**Main Contractor:** Hochtief AG
**Other Consultants:** Ingenieurbüro Schalm (Façade Engineer); RWDI (Wind Tunnel Testing); Gartner (Cladding Contractor)

### References & Further Reading

**Books:**

▶ Davies, C. & Lambot, I. (1997) *Commerzbank Frankfurt: Prototype for an Ecological High-Rise.* Watermark Publications: Basel.

▶ Fischer, V. (1997) *Sir Norman Foster and Partners: Commerzbank, Frankfurt am Main.* Axel Menges: Stuttgart.

▶ Gonçalves, J. (2010) *The Environmental Performance of Tall Buildings.* Earthscan Ltd.: London, pp. 240–251, 315–319.

▶ Jenkins, D. (ed.) (2004) *Norman Foster: Works 4.* Prestel: Munich, pp. 35–89.

▶ Quantrill, M. (1999) *The Norman Foster Studio: Consistency through Diversity.* E & FN Spon: London, pp. 164–169.

**Journal Articles:**

▶ Bailey, P. (1997) "Commerzbank, Frankfurt; Architects: Foster + Partners," *Arup Journal,* vol. 32, no. 2, pp. 3–12.

▶ Gonçalves, J. & Bode, K. (2011) "The importance of real life data to support environmental claims for tall buildings," *CTBUH Journal,* vol. 2, pp. 24–29.

▶ Gonçalves, J. & Bode, K. (2010) "Up in the air," *CIBSE Journal,* December, pp. 32–34.

▶ Pepchinski, M. (1998) "With its naturally ventilated skin and gardens in the sky, Foster + Partners' Commerzbank reinvents the skyscraper," *Architectural Record,* vol. 186, no. 1, pp. 68–79.

## Project Data:

**Year of Completion**
▸ 1998

**Height**
▸ 119 meters

**Stories**
▸ 23

**Gross Area of Tower**
▸ 53,068 square meters

**Building Function**
▸ Educational

**Structural Material**
▸ Concrete

**Plan Depth**
▸ 20 meters (from core)

**Location of Plant Floors:**
▸ B4, 18

## Ventilation Overview:

**Ventilation Type**
▸ Mixed-Mode:
Complementary-Changeover

**Natural Ventilation Strategy**
▸ Cross and Stack Ventilation
(connected internal spaces)

**Design Strategies**
▸ Ventilation "Wind Core" (central
escalator void)
▸ "Wind Floor" over central void
▸ Innovative window openings in
lecture rooms.

**Double-Skin Façade Cavity:**
▸ None

**Approximate Percentage of Year Natural
Ventilation can be Utilized:**
▸ 29%

**Percentage of Annual Energy Savings for
Heating and Cooling:**
▸ 55% compared to a fully air-
conditioned office building in Japan
(measured)

**Typical Annual Energy Consumption
(Heating/Cooling):**
▸ 166 kWh/m$^2$ (measured)

# Liberty Tower of Meiji University Tokyo, Japan

## Climate

The city of Tokyo is located in a temperate climate characterized by hot, humid summers and generally mild winters with cold spells. Summers are typically hot, hovering around 30 °C during the daytime with a night-time low of 23 °C. During the winter season, January is characteristically the coolest month, with daytime temperatures around 10 °C, which at night drops to around 3 °C with only an occasional dusting of snow. Although it rains most of the year, the wettest months are June (considered the rainy season) and September (typhoon season) which can create extra humidity (see Figure 2.3.1).

## Background

The Liberty Tower of Meiji University is located in the Surugadai district of central Tokyo. It was built as a symbol of the university, commemorating its 120th anniversary in 1998 (see Figure 2.3.2). The tower houses teaching and lecture spaces from floors 1–17. A graduate school is located on floors 19–23. Three subterranean levels contain a library and parking spaces below the main entrance hall to the building. The building is rectangular in plan, with four semi-cylindrical structures located at each corner of the building which house the staircase shafts and other service elements (see plan, Figure 2.3.3). The majority of the lecture rooms are located on the southeastern façade which runs parallel to the long axis of the building. The opposite northwestern façade contains the elevator cores and service areas such as toilets.

The defining feature of the plan, from a natural ventilation viewpoint, is a void which runs from floors 1–17, known as the "Wind Core." This void contains inter-floor escalators serving as a major means of vertical circulation for students (see section, Figure 2.3.4). In addition, the eighteenth floor is referred to as the "Wind Floor," which accommodates the mechanical rooms and serves as a key element in the natural ventilation strategy of the building.

## Climatic Data:[1]

**Location**
- ▸ Tokyo, Japan

**Geographic Position**
- ▸ Latitude 35° 40′ N, Longitude 139° 45′ E

**Climate Classification**
- ▸ Temperate

**Prevailing Wind Direction**
- ▸ South

**Average Wind Speed**
- ▸ 3.4 meters per second

**Mean Annual Temperature**
- ▸ 17 °C

**Average Daytime Temperature during the Hottest Months (June, July, August)**
- ▸ 29 °C

**Average Daytime Temperature during the Coldest Months (December, January, February)**
- ▸ 11 °C

**Day/Night Temperature Difference During the Hottest Months**
- ▸ 7 °C

**Mean Annual Precipitation**
- ▸ 1,529 millimeters

**Average Relative Humidity**
- ▸ 72% (hottest months); 50% (coldest months)

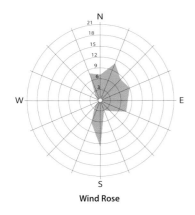

**Wind Rose**

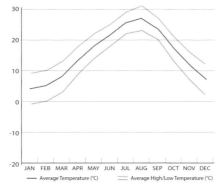

**Average Annual Temperature Profile (°C)**
— Average Temperature (°C)   ‑‑‑ Average High/Low Temperature (°C)

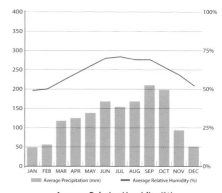

**Average Relative Humidity (%) and Average Annual Rainfall**
▮ Average Precipitation (mm)   — Average Relative Humidity (%)

▲ Figure 2.3.1: Climate profiles for Tokyo, Japan.[1]
◀ Figure 2.3.2: Overall view from the east. © Meiji University

---

[1] The climatic data listed for Tokyo was derived from the World Meteorological Organization (WMO) and the Japan Meteorological Agency.

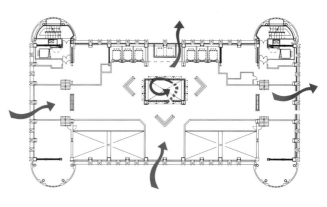

## Plan

*Fresh air enters the classrooms through openings at the base of each window and is exhausted to the escalator void through return ducts in the ceiling. The escalator void acts as a vertical "Wind Core" through stack effect. The eighteenth floor, "Wind Floor," has openings on four sides, and thus exhausts air from the "Wind Core."*

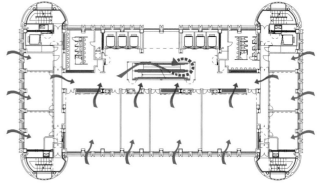

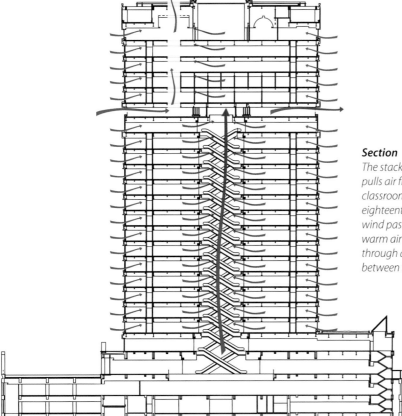

## Section

*The stack effect in the central escalator void (the "Wind Core") pulls air from the classrooms at each floor. Fresh air enters the classrooms through openings at the base of each window. The eighteenth floor, "Wind Floor" has openings on four sides. As wind passes through the Wind Floor it creates suction, pulling warm air from the escalator void. A similar effect is achieved through an atrium in the upper section of the building, between floors 19–23.*

▲ Figure 2.3.3: Eighteenth floor "Wind Floor" (top); typical floor plan (bottom).
◄ Figure 2.3.4: Building section. (Base drawings © Nikken Sekkei)

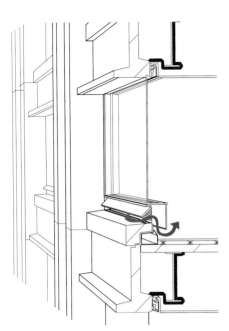

▲ Figure 2.3.5: Detailed window section showing ventilation opening. © Nikken Sekkei

▲ Figure 2.3.6: Typical classroom space, with exhaust vents visible in the ceiling (used for both forced air and natural ventilation return). © Kawasumi Kobayashi Kenji Photograph

## Natural Ventilation Strategy

The overall ventilation strategy relies on both wind and stack effect to drive air in and out of the building. The occupied rooms along the perimeter of the building feature single-glazed fixed windows with automatically controlled ventilation openings at the bottom for air entry. These ventilation openings are installed inside the window counter units (see Figure 2.3.5), designed to minimize outdoor noise and allow for fresh air intake even when blinds are closed during lectures using audiovisual equipment.

As air enters the building through the window openings and flows across the occupied spaces and lecture rooms, it is exhausted to the escalator void through return ducts in the ceiling. These ducts exhaust air in both natural and mechanical ventilation modes (see Figure 2.3.6). The escalator void acts as a vertical Wind Core by utilizing stack effect to pull the return air from the lecture rooms and exhausts it out of

the building on the eighteenth-level Wind Floor. This Wind Floor provides additional uplift for air in the Wind Core and helps induce fresh air entry through the perimeter windows on floors 1–17.

The Wind Floor is partially open to the exterior via openings on all four faces of the façade (see Figure 2.3.7), creating four wind paths and maintaining a steady flow of air through the building regardless of the prevailing wind direction. Three V-shaped glass screens, known as "Wind Fences," are placed a short distance from the top of the Wind Core (see Figures 2.3.3 & 2.3.8). These Wind Fences prevent outdoor air that is flowing through the Wind Floor from disrupting the flow of exhaust air through the top of the Wind Core. These V-shaped "deflectors" essentially ensure that exhausted air coming up the Wind Core can flow into the predominant airflow across the Wind Floor.

The graduate school in the upper section of the building, between floors 19–23 (i.e., above the Wind Floor)

**The escalator void, or Wind Core as it is known, runs from floors 1 to 17 and, through stack effect, induces airflow from the perimeter windows.**

▲ Figure 2.3.7: The large openings to the Wind Floor on the eighteenth level are clearly visible on the building's façade. © Meiji University

▲ Figure 2.3.8: View of the exhaust openings at the top of the escalator void on the Wind Floor, and one of the three Wind Fences in the foreground which protect the openings from cross winds in the Wind Floor. © Kawasumi Kobayashi Kenji Photograph

are connected by an atrium near the elevators which exhausts at the top level for similar effect.

## Mixed-Mode Strategy

The Liberty tower has a Complementary-Changeover system which switches between natural and mechanical ventilation on a seasonal or daily basis. The mechanical system comprises a single duct VAV system with two air-handling units located on every floor. The mixed-mode system thus combines passive wind-induced natural ventilation with a minimal active mechanical air-conditioning system. The Wind Core/escalator void also serves as the return path duct when mechanical systems/conditioned air are used. The air outlets on the eighteenth-level Wind Floor also serve as air exhausts in both natural and mechanical ventilation modes.

The natural ventilation mode is inactive during the heating season (i.e., winter) when the temperature of the external cold air flowing directly through the occupied spaces would be too low. At that time, the building is heated by the HVAC system at a constant supply temperature. Natural ventilation is more effective during the shoulder seasons (spring and autumn) and in summer when the outside weather conditions allow. The building management system (BMS) will determine to use a mixed or fully air-conditioned system based on daytime temperatures, wind speed, and rainfall.

## Interface with the Central Building Management System

The building has a central BMS which automatically controls the operation of perimeter ventilation openings on typical floors, and the exhaust air openings on the eighteenth-level Wind Floor. The operation of these openings depends on dry-bulb temperature, humidity, wind speed, and precipitation values, measured by sensors and meters placed on the roof of the tower. All ventilation openings are closed at high external temperatures, if it is raining and/or when the wind speed, is higher than 10 m/s. Night-purging ventilation, during the non-occupied hours of the cooling season, is automatically activated when external temperatures fall between 10 °C and 28 °C. Daytime cooling is automatically activated if the external temperature is under 22 °C and over 15 °C. Each inlet and outlet are opened automatically with monitoring of the outside temperature and the inside room temperature. The windows can also be manually controlled by occupants during the overtime period, on holidays and in case of emergency.

## Other Sustainable Design Elements

The façade of the tower features recessed glass panels and horizontal sun-shading devices integrated within the window casement. Solar control is also achieved with internal blinds limiting the intensity of direct solar radiation. Lighting is controlled by sensors which optimize the use of natural daylighting. In addition, building fabric/mass cooling is achieved by ventilating the building at night to reduce peak room temperatures during the day. Indoor air quality is also monitored using $CO_2$ sensors to control the air volume of mixed outdoor fresh air.

## Analysis – Performance Data

CFD simulations were performed to test the effectiveness of various components used to induce natural ventilation in the building. The simulations of airflow through the Wind Floor

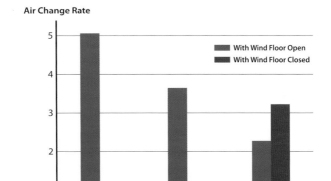

**Air Change Rate**

▲ Figure 2.3.9: BMS performance data showing air change rates per hour with the Wind Floor open and closed at various times and locations in the building. (Source: interpreted from Kato & Chikamoto 2002 p. 19)

showed that external air flows across the floor at a relatively steady speed without declining, despite the various obstacles on that floor (for example, the Wind Fences). The CFD results also revealed that the Wind Floor increased airflow rates in the occupied spaces of the tower by approximately 40 percent, thus indicating the effectiveness of the natural ventilation strategy in inducing air intake via perimeter openings on each floor (Kato & Chikamoto 2002).

In addition to carrying out CFD simulations, the building was monitored by the BMS in order to obtain actual performance data during occupancy. Measurements were automatically recorded every ten minutes from over 2,000 sensors in the building (which recorded interior and exterior temperature, humidity, air velocity values, and energy consumption rates for building equipment). These on-site measurements supported the findings of the CFD analysis in most cases, specifically that the Wind Floor increased airflow rates in the occupied spaces of the building irrespective of predominant wind direction (see Figure 2.3.9).

**The eighteenth-level Wind Floor produces additional uplift for air in the wind core and uses Wind Fences to prevent disruption of airflow.**

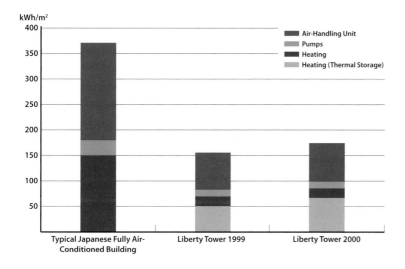

kWh/m²

Legend:
- Air-Handling Unit
- Pumps
- Heating
- Heating (Thermal Storage)

Typical Japanese Fully Air-Conditioned Building | Liberty Tower 1999 | Liberty Tower 2000

▲ Figure 2.3.10: Annual energy consumption for heating and cooling Liberty Tower, compared to a typical fully air-conditioned building in Japan. (Source: interpreted from Kato & Chikamoto 2002 p. 25)

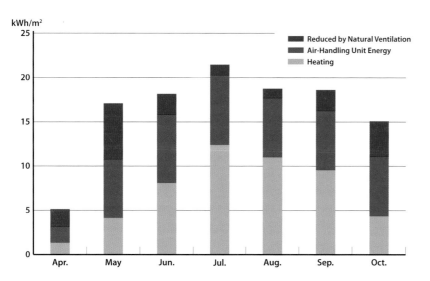

kWh/m²

Legend:
- Reduced by Natural Ventilation
- Air-Handling Unit Energy
- Heating

Apr. | May | Jun. | Jul. | Aug. | Sep. | Oct.

▲ Figure 2.3.11: Average monthly energy consumption for cooling load. (Note: monthly cooling loads are added together to create the annual cooling loads seen in Figure 2.3.10 (Source: Kato & Chikamoto 2002 p. 24)

**Fresh air inlets, which are located in the bottom part of perimeter windows, may lead to occupant discomfort, especially for those sitting near windows on the top floors of the building.**

One example showed the airflow rate in a north-facing room on the sixth floor increased from 0.5–0.7 air changes per hour with the Wind Floor closed, to 4.6–5.5 air changes per hour with the Wind Floor opened (Kato & Chikamoto 2002). However, there were cases in some locations during some wind speed/direction conditions where a drop in air changes actually occurred. For example, an eleventh-floor north-facing room showed a decrease from 3.2 air changes per hour to 2.1–2.4 air changes per hour when the Wind Floor was opened. However, these instances have not comprehensively

had a negative impact on the building's performance.

The overall performance of the building has achieved significant savings in energy consumption. The use of natural ventilation has reduced the cooling energy by typically 40 percent in the month of April, 62 percent in November and 6 percent in July (Kato & Chikamoto 2002). During monitoring in the first few years of operation, Liberty Tower had achieved a 55 percent annual energy saving for heating and cooling, compared to a fully air-conditioned building in Japan, according to local Japanese codes based on the Coefficient of Energy consumption for air-conditioning and for ventilation. The total annual heating and cooling energy was 166 kWh/m² (see Figure 2.3.10), with the total annual energy for cooling alone being 95 kWh/m², which was a 17 percent reduction from a fully air-conditioned Japanese building according to code at the time of completion in 1998 (see Figure 2.3.11) (Kato & Chikamoto 2002).

### Analysis – Strengths

▸ The building presents a good example of how a high-density-occupancy space (i.e., classrooms and lecture theaters) can be naturally ventilated, and how an escalator void can transcend its function as purely a circulation element, to be exploited as a ventilation shaft/extraction chimney in a tall building.

▸ The Wind Floor enhances the stack effect in the escalator void, and therefore induces the entry and flow of fresh air across the occupied spaces below. In this sense, the use of the combined forces of wind and buoyancy enhances natural ventilation.

- The exhaust openings on all perimeter faces of the Wind Floor are designed to account for different wind paths, thus maintaining a steady driving force for the stack effect and for air exhaust regardless of changes in wind direction.

- In a conventional air-conditioned tall building, several floors are typically needed to accommodate the mechanical HVAC equipment. The Wind Floor featured in this building presents a good example of how one floor can be dedicated to accommodating both the mechanical equipment and supporting natural ventilation in a mixed-mode high-rise building.

- The design of the exterior ventilation openings allows fresh air intake even when blinds are closed due to lectures using audiovisual equipment.

## Analysis – Considerations

The following areas could be causes for concern if adopting similar strategies in other buildings and should therefore be considered:

- With fresh air inlets located in the bottom part of perimeter windows, there is a risk that the direct introduction of air at this level may lead to occupant discomfort, especially for those sitting near windows on the top floors of the building (where wind speed is higher).

- The use of both a Wind Floor and a Wind Core might lead to the development of an extreme stack effect in the escalator core, inducing airflow rates in excess of those comfortable for normal occupancy.

- The motorized ventilation openings are fully controlled by the BMS. The inability of occupants to control their own environment might limit their comfort range and result in user dissatisfaction.

- Another concern is the potential pressure loss between the occupied rooms and the Wind Core. Return air from the rooms, including both natural ventilation and conditioned air, are led through return path ducts to the Wind Core. The size limitation of these ducts, including smoke and fire dampers, could lead to significant pressure loss and a reduced airflow rate (Kato & Chikamoto 2002).

- The complete free flow of air around the building, including an 18-story void, creates significant hazards for fire and smoke control that need to be addressed. In the case of Liberty Tower, automatic fireproof and smoke proof shutters are lowered around the eighteenth-floor atrium openings in emergency events.

## Project Team

**Owner/Developer:** Meiji University
**Design Architect:** Nikken Sekkei Ltd.
**Structural Engineer:** Nikken Sekkei Ltd.
**MEP Engineer:** Nikken Sekkei Ltd.
**Main Contractor:** Takenaka Corporation

**References & Further Reading**

**Conference Papers:**

- Chikamoto, T., Kato, S. & Ikaga, T. (1999) "Hybrid air-conditioning system at Liberty Tower of Meiji University," paper presented at 1999 IEA Energy Conservation in Buildings & Community Systems Annex 35 Conference on Hybrid Ventilation, Sydney, Australia, 28 September–1 October.

- Kato, S. & Chikamoto, T. (2002) "Pilot study report: the Liberty Tower of Meiji University," paper presented at 2002 IEA Energy Conservation in Buildings & Community Systems Annex 35 Conference on Hybrid Ventilation, Montreal, Canada, 13–17 May.

## Project Data:

**Year of Completion**
- ▸ 1998

**Height**
- ▸ 94 meters

**Stories**
- ▸ 21

**Gross Area of Tower**
- ▸ 13,777 square meters

**Building Function**
- ▸ Office

**Structural Material**
- ▸ Concrete

**Plan Depth**
- ▸ 14 meters (from core)

**Location of Plant Floors:**
- ▸ Semi-Decentralized/Every Floor

## Ventilation Overview:

**Ventilation Type**
- ▸ Mixed-Mode:
  Complementary-Alternate

**Natural Ventilation Strategy**
- ▸ Wind-Driven Cross-Ventilation

**Design Strategies**
- ▸ "Wing Walls" which capture a wider
  range of wind directions

**Double-Skin Façade Cavity:**
- ▸ None

**Approximate Percentage of Year Natural
Ventilation can be Utilized:**
- ▸ 0–100% (depending on tenant)

**Percentage of Annual Energy Savings for
Heating and Cooling:**
- ▸ 25% compared to a fully air-
  conditioned office building in
  Malaysia (measured)

**Typical Annual Energy Consumption
(Heating/Cooling):**
- ▸ 180 kWh/m² (measured)

# Case Study 2.4
# Menara UMNO Penang, Malaysia

## Climate

The tropical climate of Penang endures consistently hot weather throughout the year, seeing temperatures hover around 31 °C. Occasional rain showers can occur at any time of year. However, extreme rainy weather combined with high winds is most likely to occur during April and May, and again from August through October when the southwesterly monsoon season arrives. The most torrential downpours tend to be short-lived and sunshine is never far behind. Humidity is also fairly high throughout the year (see Figure 2.4.1).

## Background

The Menara UMNO is an office building located in the center of Penang. The building is situated on a restricted urban infill site but suffers from no interference from other high-rise buildings (see Figure 2.4.2). The form and orientation of the tower are a result of site limitations. Since the tower could not be orientated to totally maximize

prevailing winds, a series of wind "Wing Walls" were developed (see "Natural Ventilation Strategy").

The building accommodates 14 floors of office space, atop a seven-story podium. The building is almost rectangular in plan, with a gentle curve along the southwest corner of the tower (see plan and section, Figures 2.4.3 & 2.4.4). The building's long axis is oriented along the northeast–southwest direction, in line with the two prevailing wind directions. The service core (containing elevator lobbies, staircases, and toilets) is placed along the southeast façade of the building, constituting a thickly buffered party wall that shades the offices from the morning sun (see Figure 2.4.5). This configuration also allows the public and circulation areas to receive natural light and ventilation, thus reducing the energy needed to operate these spaces. The remainder of each floor area in the tower is devoted to open-plan offices. The floor plan depth is approximately 14 meters, excluding the service core. Multiple balconies are incorporated within the tower and are intended to

## Climatic Data:[1]

**Location**
▸ Penang, Malaysia

**Geographic Position**
▸ Latitude 5° 18′ N, Longitude 100° 16′ E

**Climate Classification**
▸ Tropical

**Prevailing Wind Direction**
▸ South-southwest

**Average Wind Speed**
▸ 2.6 meters per second

**Mean Annual Temperature**
▸ 28 °C

**Average Daytime Temperature during the Hottest Months (February, March, April)**
▸ 32 °C

**Average Daytime Temperature during the Coldest Months (September, October, November)**
▸ 31 °C

**Day/Night Temperature Difference During the Hottest Months**
▸ 8 °C

**Mean Annual Precipitation**
▸ 2,398 millimeters

**Average Relative Humidity**
▸ 79% (hottest months); 73% (coldest months)

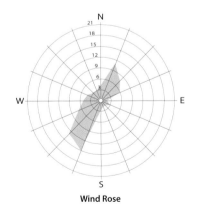

**Wind Rose**

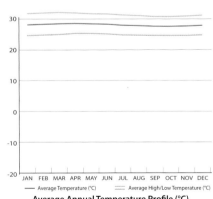

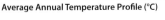

**Average Annual Temperature Profile (°C)**

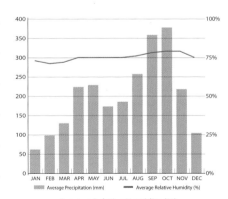

**Average Relative Humidity (%) and Average Annual Rainfall**

▲ Figure 2.4.1: Climate profiles for Penang, Malaysia.[1]
◀ Figure 2.4.2: Overall view from the west. © T.R. Hamzah and Yeang

[1] The climatic data listed for Penang was derived from the World Meteorological Organization (WMO) and Malaysia Meteorological Department.

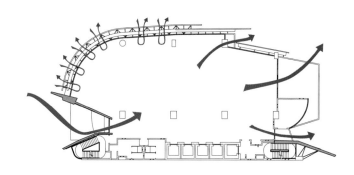

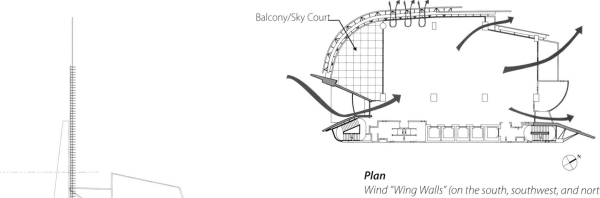

Balcony/Sky Court

## Plan

*Wind "Wing Walls" (on the south, southwest, and northeast façades – highlighted in green) are used to capture a wide range of wind directions. Windows adjacent to the Wing Walls and balconies allow for cross-ventilation at each floor. Additionally, windows along the west and southwest curving façade can be opened for direct ventilation.*

## Section

*Cross-ventilation with a high air change rate is used to create comfort ventilation. The Wing Walls increase pressure on the windward façade, increasing airflow rates into the interior. Fresh air enters the building through windows located adjacent to the Wing Wall facing the prevailing wind and exhausts through windows adjacent to the Wing Wall on the leeward side of the tower.*

▲ Figure 2.4.3: Typical office floor plan (top); typical office floor plan with balcony/sky court (bottom).
◀ Figure 2.4.4: Building section. (Base drawings © T.R. Hamzah and Yeang)

▲ Figure 2.4.5: The service core is placed along the southeast façade. © T.R. Hamzah and Yeang

▲ Figure 2.4.6: View from one of the balconies. © T.R. Hamzah and Yeang

**Wing Walls were developed to maximize the use of the prevailing winds by capturing a wider range of wind directions and increase the airflow rate into the interior.**

provide solar shading and a social space where occupants can enjoy the cooling breezes of Penang (see Figure 2.4.6).

## Natural Ventilation Strategy

The warm, humid climate of Malaysia necessitates the use of air-conditioning in most office buildings. However, poor rental rates in Penang at the time of Menara UMNO's creation did not justify the installation of a central air-conditioning system due to its economic impracticality. As a result, the tower was initially designed for tenants to install individual split-unit air-conditioning with natural ventilation conceived as a backup system in the incident of a power failure. Prior to construction, a central air-conditioning system was designed and subsequently installed. Since the original natural ventilation system was conceived early in the design process, all office floors can be naturally ventilated if external weather conditions are suitable. However, currently only the elevator lobbies are naturally ventilated.

Due to site limitations and the orientation of the building, a series of wind "Wing Walls" were developed to maximize the use of the prevailing winds. The Wing Wall device is a shortened wall that runs vertically up the building in order to capture a wider range of wind directions and increase the airflow rate into the interior. When the prevailing wind is at an oblique angle to the window inlet, the cross-ventilation strategy could not solely rely on negative pressure and suction created at the leeward side of the building. Instead, the Wing Walls create positive pressure on the windward side of the building needed to draw fresh air into the office spaces. Although fairly common in low-rise construction, the UMNO tower was the first high-rise office building to employ a wind Wing Wall system for the purpose of natural ventilation.

The objective of the natural ventilation solution was to generate a high air change rate which would achieve comfort conditions through air movement around occupants directly, and

through temperature control. One way to improve occupant comfort with natural ventilation is through comfort ventilation, also referred to a psychological cooling. Psychological cooling is a form of passive low-energy cooling through a direct physiological effect on the occupants in which higher indoor air speeds make occupants feel cooler. Thus the building exploits wind-induced natural ventilation not solely for the purpose of air-displacement and fresh air supply, but also for internal comfort conditions. The vertical Wing Walls run up the full height of the building and protrude from the northeast, south and southwest elevations (see Figure 2.4.7).

The overall natural ventilation strategy in the building is primarily dependant on wind-driven forces to channel air across each floor plate (cross-ventilation on a floor-by-floor basis). The main openings (windows and balcony doors) are located on the southwest and northeast elevations, which induce cross-ventilation in the direction of prevailing winds. Even if interior

▲ Figure 2.4.7: Wing Walls on the southwest. © T.R. Hamzah and Yeang

## Mixed-Mode Strategy

Menara UMNO was originally designed as a Contingency Mixed-Mode building to be naturally ventilated for up to 100 percent of the year with the provision for tenets to install individual air-conditioning units. Subsequently, a central air-conditioning system was installed during construction and the tower operates as a Complementary-Changeover building. A user-controlled system allows occupant to choose between natural ventilation or mechanical ventilation modes, enabling them to run in either mode indefinitely.

On each floor, air-handling units (AHUs) located in plant rooms along the northeast façade draw in fresh air through intake openings on the façade. The AHUs have a cooling capacity of 100 kW and are connected through a common riser to three cooling towers on the roof. The ductwork and diffusers contained in the suspended ceiling supply fresh air to the offices, while return vents and ceiling voids allow air to return to the AHUs.

## Interface with the Central Building Management System

The building does not have a central building management system (BMS) to control the operation of the windows or the amount of airflow into the interior. However, the active and passive systems in the building have independent controls which can be operated by the users. The controls for the active AHU are on a floor-by-floor basis, as opposed to being centralized. For the passive system, occupants have access to windows that can be opened in most workspaces. It is up to the tenants to decide whether to manually use natural ventilation or use the automatic control system for air-conditioning.

**Multiple balconies are incorporated within the tower and are intended to provide solar shading and a social space where occupants can enjoy the cooling breezes of Penang.**

partitions are erected between the office space and elevator lobbies, the elevator lobbies can remain naturally ventilated. Other windows are located along the northwest façade for user-controlled direct ventilation. As air is funneled toward the building from the windward side, it is directed to a zone with special balconies that serve as pockets with "airlocks." These airlocks have adjustable doors and windows that can be manually opened or closed to control the rate and the distribution of natural ventilation within the space. Occasional balconies cut out at an angle on the west-facing corner provide further natural ventilation and solar shading for the tower.

## Other Sustainable Design Elements

Achieving occupant comfort is dependent on both the utilization of airflow and the minimization of solar gain. Without minimization techniques, the increase in solar gain would overwhelm the comfort gains of natural ventilation.

Three different techniques were used to minimize solar heat gains while retaining sufficient daylight and reasonable views. First, placing the opaque concrete walls of the service core along the southeast façade and the AHU plant rooms on the northeast façade reduces solar heat gain from the low morning sun. Second, the glazed northwest and southwest façades are protected from afternoon sun by sky court balconies and perforated, horizontal aluminum screens acting as solar shading devices. The screens are deeper and more opaque toward the west and more transparent toward the north. Last, the rooftop of the building, which would receive most of the solar heat gain at this latitude, is shaded by a large curved roof canopy and the upward extension of the elongated service core.

## Analysis – Performance Data

During the design stage, CFD simulations were conducted to test the effectiveness of various ventilation strategies. These studies concluded that comfort ventilation cannot be achieved solely based on buoyancy-induced ventilation due to low floor-to-floor heights and the strategy to ventilate on a floor-by-floor basis. The CFD studies then examined the effectiveness of the Wing Walls and the thermal performance of the building in reducing interior temperatures. Wind flow around

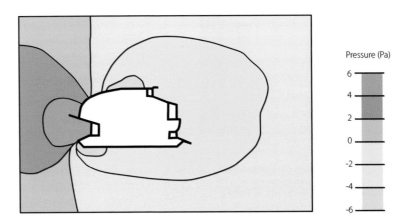

Pressure (Pa)

▲ Figure 2.4.8: CFD simulation of air pressure contours which show wind flow around the building in plan.
© T.R. Hamzah and Yeang

the building was simulated using the airflow model DFS-AIR to obtain values of the surface pressures at each window and door opening. The test controls were a southwest prevailing wind at a speed of 2.5 m/s at a height of 10 meters from ground level and also accounted for wind speed increasing at higher elevations. The results confirmed the increase in pressure at the Wing Walls (see Figure 2.4.8).

CFD studies were also used in determining interior air change rates. For a typical density in an office building, 1–2 air changes per hour may be sufficient to supply ventilation needs. However, five air changes per hour may be required to exhaust typical office heat gains (occupants, lighting and equipment), such that internal air temperature is within 1 °C of the external air temperature. Further simulations calculated internal air change rates under various wind speeds. External factors included a calm day with no wind (which determined airflow rates generated by thermal buoyancy alone) and three wind conditions, 1.0, 2.5, and 5.0 m/s, with a range of window opening configurations. The

test control used a typical external air temperature of 30 °C with internal heat gains assumed to be 35 W/m². The test resulted in a range of air changes per hour from 1 to 33.8. Such variations in ventilation rates necessitate the use of adjustable openings which can be closed down to a minimum, especially in the windward direction.

As mentioned previously, solar heat gain is a great concern of building design in tropical climates such as Malaysia. A study was done which illustrates the impact of shading (see Figure 2.4.9). The total energy load for cooling Menara UMNO is 180 kWh/m². This is a 7 percent reduction from the energy load if the building did not utilize shading devices. Further, the combination of natural ventilation and shading results in a 25 percent reduction as compared to a typical air-conditioned building in Malaysia (Jahnkassim & Ip 2006).

A post-occupancy analysis was conducted in 1998 which measured thermal comfort using Fanger's Predicted Mean Vote (PMV).[2] The PMV

---

[2] Fanger's Predicted Mean Vote (PMV) is a comfort equation created by P. O. Fanger which is used to evaluate occupant comfort conditions.

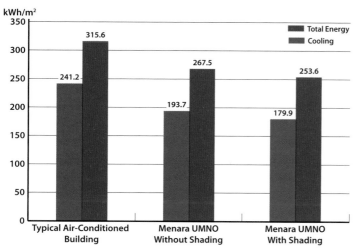

▲ Figure 2.4.9: Comparison of energy use in Menara UMNO compared to a typical air-conditioned building in the region (showing a 25 percent reduction), as well as the positive impact of shading devices (showing a 7 percent reduction over the same design with no shading). (Source: Jahnkassim & Ip 2006)

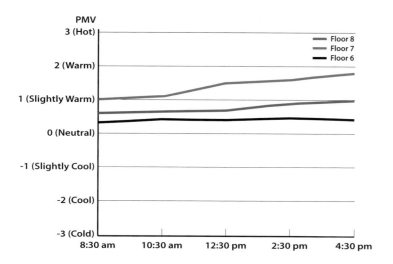

▲ Figure 2.4.10: A post-occupancy analysis was conducted in 1998 which measured thermal comfort using Fanger's PMV, which revealed that building tenants tended to feel slightly warm while in the building. (Source: Jahnkassim 2004)

**Comfort ventilation is a form of passive low-energy cooling through a direct physiological effect on the occupants in which higher indoor air speeds make occupants feel cooler.**

model combines four physical variables (air temperature, air velocity, mean radiant temperature, and relative humidity) and two personal variables (clothing insulation and activity level) into an index that predicts thermal comfort. On a seven-point scale, 0 represents neutral (neither hot nor cold), positive 3 represents feeling hot, and negative 3 represents feeling cold. The study revealed occupants on floors 6, 7, and 8 tended to feel slightly warm, with an average scoring of 0.9 PMV (see Figure 2.4.10) (Jahnkassim 2004).

However, another post-occupancy study was conducted in July 2004. This study consisted of a questionnaire distributed to occupants in order to record the perceptions of the environmental qualities of their work space (Ismail & Sibley 2006). The survey addressed natural air ventilation, thermal comfort, and work environment satisfaction. The study revealed that natural ventilation had an 83 percent positive response. For thermal comfort, there was only a 57 percent positive response, with negative feedback resulting from questions assessing the quality of temperature at work stations, elevator lobbies, and restrooms. Despite the results of the thermal comfort survey, none of the respondents were dissatisfied with their overall working environment.

### Analysis – Strengths

▸ The orientation of the Wing Walls along the prevailing wind axes allow air to be funneled into the interior.

▸ Considering the relatively low average wind speeds in Penang (2.6 m/s), the Wing Wall becomes a crucial element of the design as it channels air into the interior at higher rates/speeds, thus serving to achieve comfort ventilation in Penang's hot and humid climate.

▸ The Wing Walls direct a greater range of prevailing winds into the building, minimizing the risks associated with natural ventilation due to changes in wind directions.

▸ The location of the service core on the eastern façade allows it to act as a solar buffer on the wide axis of the building.

## Analysis – Considerations

The following areas could be causes for concern if adopting similar strategies in other buildings and should therefore be considered:

▸ Since the ventilation concept mainly relies on wind-driven air, low airflow rates at about 1 ac/h would be attained if the wind speeds are low or the windows are not fully opened. This causes the internal/external air temperature difference to be no greater than 1.5 °C. Regarding the low temperature difference and low ceiling height, stack effect alone would not provide a significant comfort cooling effect. Moreover, the absence of a central computerized BMS-controlling the windows might cause problems in attaining the required airflow rates needed for comfort ventilation in this humid climate.

▸ At the other extreme, with high · wind speeds, an air change rate of 28 or higher is too high for comfort and would cause problems such as papers moving or flying around. Partially closing the windows would provide a solution. However, without BMS controlled windows, it is difficult to anticipate how the occupants will control their environment, i.e., closing windows and turning on air-conditioning as opposed to still utilizing natural ventilation by partially closing the windows.

▸ As mentioned previously, CFD simulations resulted in airflow rates as low as 1 ac/h under buoyancy alone conditions and between 6.3 and 28 ac/h with high external wind speeds. Given the design of the building, these substantial variations diminish the possibility of a tall building that relies exclusively on natural ventilation without sacrificing occupant comfort.

▸ Since high humidity levels were not accounted for, the sole use of natural ventilation may not provide optimal occupancy comfort.

▸ Under tropical conditions, the effect of wind-driven rain under monsoon conditions is likely to have a major negative impact.

▸ The location of the service core along the edge of the building has resulted in a greater plan depth along the main wind flow axis (approximately 14 meters). This might have adverse effects on natural daylighting, especially for workers who are located at a far distance from windows and ventilation openings. It also presents a completely blank façade to the city on one side (see Figure 2.4.5).

▸ The development of future adjacent high-rises may negatively affect wind flow and speeds and thus the natural ventilation strategy.

## Project Team:

**Owner/Developer:** United Malays National Organisation (UMNO)
**Design Architect:** T.R. Hamzah and Yeang
**Structural Engineer:** Tahir Wong Sdn Bhd
**MEP Engineer:** Ranhill Bersekutu
**Main Contractor:** JDC Corporation Sdn Bhd
**Other Consultants:** University of Cardiff (Natural Ventilation Modeling)

**References & Further Reading**

**Books:**

▸ Ismail, L. H. & Sibley, M. (2006) "Bioclimatic performance of high-rise office buildings: a case study in Penang," in Compagnon, C. Haefeli, P. & Weber, W. (eds.) *Proceedings of PLEA 23rd International Conference*. Imprimerie St-Paul: Fribourg, Switzerland, pp. 981–986.

▸ Jahnkassim, P. S. & Ip, K. (2000) "Energy and occupant impacts of bioclimatic high-rises in a tropical climate," in Steemers, K. & Yannas, S. (eds.) *Proceedings of PLEA 17th International Conference*. James and James (Science Publishers) Ltd.: London, pp. 249–250.

▸ Powell, R. (1999) *Rethinking the Skyscraper: The Complete Architecture of Ken Yeang*. Thames and Hudson: London, pp. 82–91.

▸ Richards, I. (2001) *T. R. Hamzah & Yeang: Ecology of the Sky*. The Images Publishing Group: Mulgrave, Australia, pp. 170–181.

**Journal Articles:**

▸ Chye, L.P. (1998) "Menara UMNO, Jalan Macalister, Pulau Pinang," *Architecture Malaysia*, vol. 10, no. 1, pp. 32–35.

▸ Powell, R. (1998) "Vertical aspirations – Menara UMNO: Penang, Malaysia; architects: T. R. Hamzah & Yeang," *Singapore Architect*, vol. 200, pp. 66–71.

**Conference Papers:**

▸ Jahnkassim, P. S. & Ip, K. (2006) "Linking bioclimatic theory and environmental performance in its climatic and cultural context – an analysis into the tropical high-rises of Ken Yeang," paper presented at PLEA 2006 Conference on Passive and Low Energy Architecture, Geneva, Switzerland, 6–8 September.

▸ Jones, P. J. & Yeang, K. (1999) "The use of the wind wing-wall as a device for low-energy passive comfort cooling in a high-rise tower in the warm mid tropics," paper presented at PLEA 16th International Conference, Brisbane, Australia, 22–24 September.

**Theses:**

▸ Jankassim, P. S. (2004) "The bioclimatic skyscraper: a critical analysis of the theories and designs of Ken Yeang," PhD thesis, University of Brighton.

## Project Data:

**Year of Completion**
▸ 1999

**Height**
▸ 82 meters

**Stories**
▸ 20

**Gross Area of Tower**
▸ 13,563 square meters

**Building Function**
▸ Office

**Structural Material**
▸ Concrete

**Plan Depth**
▸ 24 meters (between façades)

**Location of Plant Floors:**
▸ 20

## Ventilation Overview:

**Ventilation Type**
▸ Mixed-Mode: Complementary-Changeover

**Natural Ventilation Strategy**
▸ Wind-Driven Cross-Ventilation
▸ Stack Ventilation via Vertical Shaft in Core (Ventilation Tower)

**Design Strategies**
▸ "Corridor" double-skin façade (large-volume horizontal air duct)
▸ Ventilation tower to exhaust air

**Double-Skin Façade Cavity:**
▸ Depth: 1.4 meters
▸ Horizontal Continuity: Fully Continuous (around entire perimeter)
▸ Vertical Continuity: 3.2 meters (floor-to-floor)

**Approximate Percentage of Year Natural Ventilation can be Utilized:**
▸ Unpublished

**Percentage of Annual Energy Savings for Heating and Cooling:**
▸ Unpublished

**Typical Annual Energy Consumption (Heating/Cooling):**
▸ 43 kWh/m$^2$ (heating only) (estimated)

# Deutsche Messe AG Building Hannover, Germany

## Climate

The city of Hannover is located in a temperate climate with warm summers and cold, humid winters. During the summer months, from June to August, daytime temperatures average 22 °C, with the occurrence of common afternoon thunderstorms. Winters can be cold and damp with an average high temperature of 5 °C that often approaches −1 °C in the evening (see Figure 2.5.1).

## Background

The Deutsche Messe AG is a German trade fair organization created after the Second World War. The trade fair site in Hannover is located along the River Leine and hosts numerous events annually. One such prominent event was the first world exhibition ever staged in Germany, the EXPO 2000. Completed in 1999, the 20-story Deutsche Messe AG administration building was designed for the EXPO. Due to limited site conditions, the decision was made to build vertically (see Figure 2.5.2).

The building, a square office tower (24 × 24 meters in plan) has two cores set diagonally opposite each other at the northeast and southwest corners of the tower (see plan and section, Figures 2.5.3 & 2.5.4). The northeast core provides vertical circulation (containing elevators and a staircase) while the southwest core houses sanitary facilities, a secondary staircase, and a freight elevator. Office floors are divided in a variety of ways: open-plan, cellular, or a combination of both. Depending on requirements, approximately 15 cellular offices can be located next to the façade on each floor. While the central zone of each floor is used for communal purposes, it allows for great flexibility in the layout and use of the building. A three-story lobby is located below the office floors, and the top floor is dedicated to the mechanical plant.

## Natural Ventilation Strategy

The exploitation of a double-skin façade is a key aspect of the tower's natural ventilation strategy. The key concept of this was to manage openings on the

## Climatic Data:[1]

**Location**
▸ Hannover, Germany

**Geographic Position**
▸ Latitude 52° 22′ N, Longitude 9° 43′ E

**Climate Classification**
▸ Temperate

**Prevailing Wind Direction**
▸ West

**Average Wind Speed**
▸ 2.6 meters per second

**Mean Annual Temperature**
▸ 9 °C

**Average Daytime Temperature during the Hottest Months (June, July, August)**
▸ 22 °C

**Average Daytime Temperature during the Coldest Months (December, January, February)**
▸ 5 °C

**Day/Night Temperature Difference During the Hottest Months**
▸ 10 °C

**Mean Annual Precipitation**
▸ 641 millimeters

**Average Relative Humidity**
▸ 72% (hottest months); 85% (coldest months)

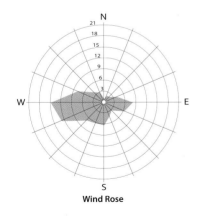

Wind Rose

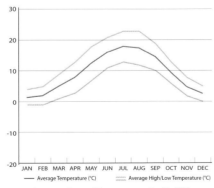

Average Annual Temperature Profile (°C)

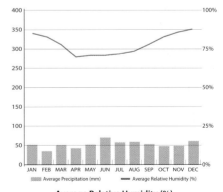

Average Relative Humidity (%) and Average Annual Rainfall

▲ Figure 2.5.1: Climate profiles for Hannover, Germany.[1]
◀ Figure 2.5.2: Overall view from the southeast. © Dieter Leistner

[1] The climatic data listed for Hannover was derived from the World Meteorological Organization (WMO) and Deutscher Wetterdienst (German Weather Service).

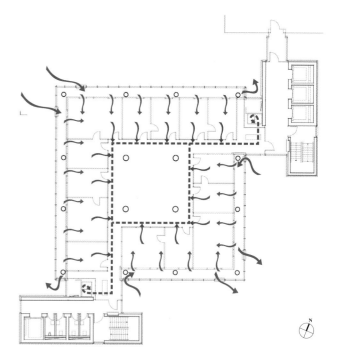

### Plan

Fresh air is introduced in the double-skin façade through eight vents running horizontally up the entire building façade. At least one sliding window is located in the internal façade of each office for natural ventilation. Stale air is exhausted via a duct in the ceiling then pulled up into a vertical shaft at each core through stack effect.

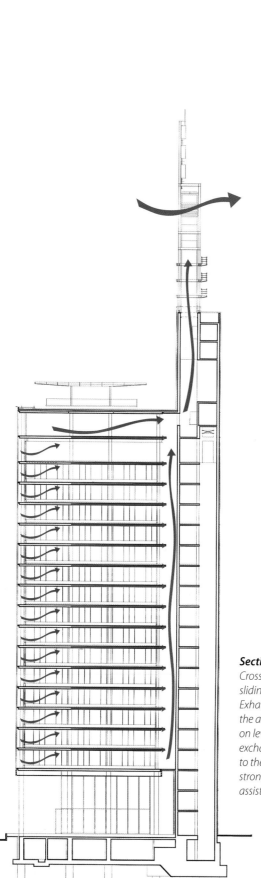

### Section

Cross-ventilation is introduced into the office spaces through sliding windows on the interior face of the double-skin façade. Exhaust air is extracted via a duct in the ceiling which transfers the air to a vertical shaft located in each core. At the plant floor on level 20, exhaust air from both cores is combined and a heat exchange unit recovers discharged energy before it is exhausted to the exterior. In addition to stack effect in each vertical shaft, strong winds at the top of the northeast core create suction to assist in the air exhaust.

▲ Figure 2.5.3: Typical office floor plan.
◀ Figure 2.5.4: Building section. (Base drawings © Herzog + Partner)

▲ Figure 2.5.5: View into the 1.4-meter-deep double-skin façade. © MoritzKorn

positive and negative pressure sides to horizontally flush out any heat gain. The double-skin façade consists of two double-glazed skins with a 1.4-meter-deep cavity. The two façade cavities, one running along the north and west façades, and a second along the south and east façades (see Figure 2.5.5) are linked by 300 mm diameter ducts in the ceiling, creating a horizontally continuous double-skin "corridor." The façade cavities are segmented at each floor by concrete slabs extending beyond the inner façade layer. The vertical segmentation of the double-skin façade creates a series of large-volume horizontal air ducts around the building. Fresh air enters the offices through operable, 1.8-meter-wide, floor-to-ceiling sliding casement windows in the inner skin. Every office space contains at least one of these openings.

The outer skin admits and exhausts fresh air into the cavity through ventilation louvers in eight locations around each floor. Four louvers are located on the north/west cavity and four on the south/east cavity. These louvers run vertically up the entire façade, though they are segmented at each floor (see Figure 2.5.6). The locations of the

louvers were determined on the basis of extensive wind-tunnel tests and simulations. Depending on pressure and temperature conditions, the louvers can be tilted at six different angles.

Exhaust air is collected on each floor through a central conduit system located at the center of the office areas and then enters one of the two vertical shafts located near the cores. Through the exploitation of stack effect, air is drawn up through these vertical shafts to the top of the building, where it is exhausted through the ventilation tower that rises approximately 30 meters above the northern core. Exhaust air is further induced by thermal up-currents created by the building's height and strong winds at the top of the tower, which induce a powerful suction effect.

## Mixed-Mode Strategy

The building employs a Complementary-Changeover system which switches between mechanical and natural ventilation on a seasonal or daily basis. When the mechanical ventilation mode is activated, fresh air

is drawn into the northern core at roof level. During winter, incoming fresh air is preheated with the aid of a rotary heat exchange unit that recovers up to 85 percent of the thermal energy discharged from exhaust air. After conditioning the air (heating/cooling) in the service floor at the top of the building, the air intake is fed into the two large shafts (one in each core) which supply each floor with a 1.5 air change rate (the requirement to maintain hygienic conditions). Concurrently, exhaust air is drawn through two shafts (which merge in the service floor) and discharged at the top of the building. The air exchange rate for each floor is controlled by electrically operated volume-flow regulators.

Supply air ducts of the mechanical ventilation system are concealed below the sill of the inner façade. A small air duct is incorporated in the apron panel below the window sill which supplies mechanical ventilation when the windows are closed (see Figure 2.5.7). These ventilation devices, which supply warm air during winter and cool air during summer, are automatically deactivated when the window is opened.

▲ Figure 2.5.6: Air intake louvers on the façade exterior. © Dieter Leistner

## Interface with the Central Building Management System

Users can choose between natural and mechanical ventilation by simply opening the sliding windows. This allows for individual control over the amount and temperature of the air supply according to personal requirements. When the windows are open, the building management system (BMS) automatically shuts off the outlets of air ducts incorporated in the apron panels of the timber casement windows.

The BMS also controls the adjustment of the air intake louvers on the outer façade. The control and operation of these louvers was developed on the basis of real-time computerized input using system data and information from weather stations, wind-tunnel tests, and analytical values. Six temperature measurement points were installed in the façade cavity to control the minimum, maximum, and estimated mean temperature values.

Various control and operation strategies are applied for the following situations: (i) seasonal differences such as selected temperature ranges relating to spring, summer, high summer, autumn, winter, and deep winter; (ii) time of day; (iii) solar heat gain in the façade; and (iv) external wind speed and direction.

**Users can choose between natural and mechanical ventilation by simply opening the sliding windows. This allows for individual control over the amount and temperature of the air supply according to personal requirements.**

## Other Sustainable Design Elements

Supplemental heating and cooling are based on thermal activation of the building's exposed concrete slabs which incorporate a water pipe system to provide further conditioning of the office spaces. In the heating season when external temperatures are below 0 °C and interior temperatures fall below 23 °C, the heating systems come into operation. Conversely, in the cooling season when exterior temperatures are above 0 °C and interior temperatures exceed 21 °C, the thermoactive slab cools the room by extremely small temperature differences (Herzog 2000). During the summer nights, the concrete structure gives off heat to internal water pipes which is circulated in contact with outside air. A cooling water temperature of up to 18 °C can be achieved.

The tower's double-skin façade forms a thermal buffer, mediating temperatures between interior and exterior and tempering fresh air before it enters the office spaces. Adjustable sunshades are installed in the façade cavity (behind the outer layer of glazing) to protect the building from solar heat gains. A heat exchange unit recovers up to 85 percent of energy from discharged exhaust air.

## Analysis – Strengths

▸ In terms of their employment for natural ventilation purposes, double-skin façades are typically used to reduce/balance wind speeds and preheat the incoming fresh air before it enters the interior during the heating season. In this building, the double-skin façade surpasses this function since the peripheral corridor of the façade is exploited as a large-volume air duct for air circulation.

- The double-skin façade acts as a "protective shield," reducing and balancing out wind forces of varying intensity that hit the building. This allows fresh air to enter the building via the intermediate façade space from all directions, irrespective of the prevailing wind speed and direction.

- The depth of the horizontally continuous double-skin façade cavity (with integral automated blinds) allows for a large volume of airflow to exhaust excess heat. This reduces the risk of overheating in the adjacent office spaces, a problem encountered with some double-skin projects.

- The building presents a good example of how the access/ service core can be exploited as a ventilation stack in a tall building. The access core transcends it's intended function to become a key element of the tower's ventilation strategy.

- The floor slabs extending beyond the inner façade layer act as fire-breaks in the double-skin façade.

- Extending the exhaust vent significantly beyond the height of the tower, and using wind to enhance stack effect, creates an effective exhaust system, which is necessary to exhaust the full-building-height vertical shafts.

## Analysis – Considerations

The following areas could be causes for concern if adopting similar strategies in other buildings and should therefore be considered:

- A deep external façade "corridor" can consume a substantial percentage of the floor area (approximately 20 percent in the case of Deutsche Messe) resulting in a loss of valuable, leasable floor area.

- A system which is dependent on occupant understanding of the systems to execute at maximum efficiency could result in problems of operation. In the case of Deutsche Messe, an occupant may open the inner casement window when ideally it should remain shut.

## Project Team

**Owner/Developer:** Deutsche Messe AG (DMAG)
**Design Architect:** Herzog + Partner
**Structural Engineer:** Sailer Stepan und Partner GmbH
**MEP Engineer:** Ingenieurbüro Hausladen GmbH; Schmidt Reuter Partner
**Other Consultants:** Design Flow Solutions (Energy Simulation)

**References & Further Reading**

**Books:**

- Herzog, T. (2000) *Sustainable Height: Deutsche Messe AG Hannover Administration Building.* Prestel Press: Munich.

**Journal Articles:**

- Ford, B. (2000) "Herzog in Hannover," *Ecotech*, vol. 1, pp. 4–7.

- Lu, W. (2007) "Special issue. Pioneering sustainable design: Thomas Herzog + Partner," *World Architecture (China)*, vol. 204, pp. 16–79.

▲ Figure 2.5.7: Office space showing mechanical supply air ducts at the base of the window casements. When windows are opened, the mechanical ventilation is automatically deactivated. © MoritzKorn

## Project Data:

**Year of Completion**
- ▸ 1999

**Height**
- ▸ 82 meters

**Stories**
- ▸ 23

**Gross Area of Tower**
- ▸ 48,000 square meters

**Building Function**
- ▸ Office

**Structural Material**
- ▸ Concrete

**Plan Depth**
- ▸ 7.2–11 meters (between façades)

**Location of Plant Floors:**
- ▸ 22


## Ventilation Overview:

**Ventilation Type**
- ▸ Mixed-Mode: Complementary-Changeover

**Natural Ventilation Strategy**
- ▸ Single-Sided Ventilation (cellular offices) Cross-Ventilation (open offices)
- ▸ Stack Ventilation via External Thermal Flue

**Design Strategies**
- ▸ Double-skin as external thermal flue
- ▸ "Wing Roof" accelerates the wind passing over the thermal flue which creates negative pressure

**Double-Skin Façade Cavity:**
- ▸ Depth: 1 meter (west façade); 200 mm (east façade)
- ▸ Horizontal Continuity: Fully Continuous (along entire length of façade)
- ▸ Vertical Continuity: Fully Continuous, approximately 67 meters (full height of façade)

**Approximate Percentage of Year Natural Ventilation can be Utilized:**
- ▸ 70%

**Percentage of Annual Energy Savings for Heating and Cooling:**
- ▸ 53% compared to a fully air-conditioned German office building (estimated)

**Typical Annual Energy Consumption (Heating/Cooling):**
- ▸ 150 kWh/m² (estimated)

# GSW Headquarters Tower Berlin, Germany

## Climate

The city of Berlin is located in a climate typified by large seasonal temperature differences, with warm, relatively humid summers and cold (at times severely cold) winters. During the summer months, from late May to August, the city stays pleasantly warm with many hours of sunshine. It is very unusual for a day to have no sunshine at all. However, July and August can also be unpredictable months with temperatures reaching 30 °C, or a sunny day quickly clouding over and turning rainy. Winters can be bitterly cold and damp, with an average high temperature of 4 °C that often approaches −2 °C in the evening. Snow is fairly common and there are often cold, clear, frosty days (see Figure 2.6.1).

## Background

The GSW Headquarters Tower forms the extension of an existing office tower built in the 1950s which was one of the first projects to be built during the reconstruction of Berlin. The design

strives to combine "found" fragments of the city and compose them into a three-dimensional composition through which the existing building is reintegrated into its context. Solar shading devices within the western façade are colored in hues of red and pink, and serve to give an identity to the tower while differentiating it from the separate elements of the new ensemble (see Figure 2.6.2). The form of the GSW Tower creates a gently curved arc in plan which has a minimum width of 7.2 meters at its ends, and a maximum width of 11 meters at the center (see plan and section, Figures 2.6.3 & 2.6.4). The tower was designed to accommodate a range of office layouts within the narrow, linear plan.

## Natural Ventilation Strategy

The double-skin façades along the east and west elevations of the building make up a key component of the natural ventilation strategy. The double-skin system of the west façade consists of an outer single-glazed weather screen and an inner double-glazed window

## Climatic Data:[1]

**Location**
- Berlin, Germany

**Geographic Position**
- Latitude 52° 30'N, Longitude 13° 25'E

**Climate Classification**
- Temperate

**Prevailing Wind Direction**
- West

**Average Wind Speed**
- 2.7 meters per second

**Mean Annual Temperature**
- 9 °C

**Average Daytime Temperature during the Hottest Months (June, July, August)**
- 23 °C

**Average Daytime Temperature during the Coldest Months (December, January, February)**
- 4 °C

**Day/Night Temperature Difference During the Hottest Months**
- 10 °C

**Mean Annual Precipitation**
- 571 millimeters

**Average Relative Humidity**
- 70% (hottest months); 85% (coldest months)

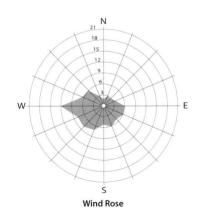

**Wind Rose**

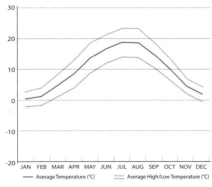

**Average Annual Temperature Profile (°C)**
— Average Temperature (°C)   — Average High/Low Temperature (°C)

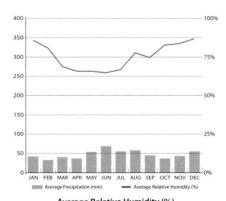

**Average Relative Humidity (%) and Average Annual Rainfall**
▬ Average Precipitation (mm)   — Average Relative Humidity (%)

▲ Figure 2.6.1: Climate profiles for Berlin, Germany.[1]
◀ Figure 2.6.2: Overall view from the west. © Annette Kisling

[1] The climatic data listed for Berlin was derived from the World Meteorological Organization (WMO) and Deutscher Wetterdienst (German Weather Service).

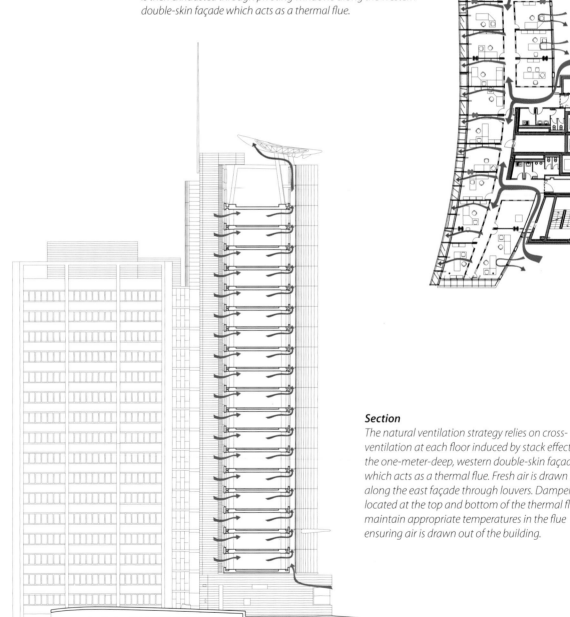

### Plan

*The natural ventilation strategy varies depending on the office configuration (see Figure 2.6.8). Double-banked offices are shown here with cellular offices on both east and west sides of the building. Fresh air is drawn into the building through pivoting panels along the east façade, directly ventilating the eastern cellular offices and drawing fresh air into the central corridor. The western cellular offices are naturally ventilated from the corridor through vents in the partitions and doors. Air is then exhausted through pivoting windows along the western double-skin façade which acts as a thermal flue.*

### Section

*The natural ventilation strategy relies on cross-ventilation at each floor induced by stack effect in the one-meter-deep, western double-skin façade which acts as a thermal flue. Fresh air is drawn in along the east façade through louvers. Dampers located at the top and bottom of the thermal flue maintain appropriate temperatures in the flue ensuring air is drawn out of the building.*

▲ Figure 2.6.3: Typical floor plan.
◀ Figure 2.6.4: Building section. (Base drawings © Sauerbruch Hutton)

enclosing a one-meter-deep cavity. The inner windows can be opened in every second bay of the façade to allow the flow of air from the offices into this intermediate space, which acts as a "thermal flue."

The double-skin system of the east façade consists of a 200 mm cavity and an outer skin which alternates between single-glazed window panels and full-floor-height, fixed louvered screens that allow fresh air into the cavity (see Figure 2.6.5). The inner layer of the east façade consists of double-glazed window panels (only operable for cleaning purposes) alternating with a pivoting panel (located directly behind the louvered screen) that allows the flow of fresh air into the interior (see Figure 2.6.6).

The flexibility in office layouts is revealed through the asymmetrical composition of the air intake louvers on the east façade (see Figure 2.6.7), where distribution expresses a direct relationship to the planning and spatial articulation of the building, as controlled and calculated air volume flow rates had to be determined to avoid excessive drafts, etc. The window openings have been laid out according to each floor plan layout. They can (and must) be adapted when plan layouts are changed (e.g., when a new tenant moves in) to guarantee the same natural air rates in all office spaces.

The natural ventilation principle in the GSW Tower can be characterized as cross-ventilation for each floor individually, whereas the overall strategy is dependent on stack ventilation to drive

▲ Figure 2.6.5: Louvered air intake openings on the east façade. © Annette Kisling

**Air is drawn from the eastern façade, across the building plan, and into the one-meter-deep continual double-skin western façade, which acts as a thermal flue.**

▲ Figure 2.6.6: Eastern double-skin façade showing the inner upper panels (aligned with the louvered intake openings) which pivot downward for air intake. © Annette Kisling

▲ Figure 2.6.7: View of the east façade, where the distribution of louvered openings expresses a direct relationship to the planning and spatial articulation of the building. © Annette Kisling

**The Wing Roof – located directly above the thermal flue top opening – uses the Venturi effect to further induce exhaust airflow in the thermal flue.**

exhaust air out of the building. Fresh air enters the building through the east façade, while exhaust air is drawn out through the "thermal flue" of the west façade, where it is eventually emitted at roof level.

Various ventilation strategies were developed for the different plan layouts. Overall, the building contains four types of office layout: (i) open-plan; (ii) double-banked with cellular offices on either side of a central corridor; (iii) single-banked with west-facing offices and a corridor along the east; and (iv) a mixture of hybrid cellular and open-plan offices with (a) east-facing cellular offices and an open-plan office space along the west side (combi-east); or (b) west-facing cellular offices and an open-plan office space along the east side (combi-west) (see Figure 2.6.8). Cross-ventilation is facilitated through the walls of cellular offices (whether glazed or solid) through ventilation

panels which were specially designed as part of the doors with a sound absorbent lining. As the air travels through each floor, acoustically baffled vent openings (ventilation panels) in the walls allow the air to enter the offices through the partitions. Once the air passes through the offices, it is drawn out through the operable windows on the west façade and into the "thermal flue," which runs continuously up the building, carrying away heat accumulated from the sun and from the offices' internal loads. Consequently, convection caused by thermal buoyancy in the western double-skin façade creates a negative pressure that induces the entry of cooler fresh air from the east side. The stack effect in the thermal flue induces cross-ventilation throughout the building when the windows on the two façades are open.

Located directly above the thermal flue opening at the top of the tower is an aerodynamic form called a "Wing Roof" (see Figure 2.6.9). It causes the wind passing over the building to accelerate, increasing the negative pressure in the flue. This phenomenon, referred to as the Venturi effect, reinforces convection and provides additional uplift for air in the flue when the wind is blowing in the right direction: west (the prevailing wind direction) or east. However, if the wind is blowing from the northerly or southerly directions, a series of fins suspended from the Wing Roof causes the wind to eddy. This prevents the risk of a positive pressure build up and the occurrence of down currents over the outlet of the thermal flue.

## Mixed-Mode Strategy

The GSW Headquarters is designed as a Complementary-Changeover building. A mechanical ventilation system was incorporated to provide comfort during seasonal weather extremes, when the external air is too hot or too cold to naturally ventilate the building. During the cooling season, no refrigeration system is used in order to maintain the concept of an environmentally responsive building. Alternatively, cooling is based on spray coolers and desiccant thermal wheels (which dehumidify the air stream). Desiccant thermal wheels are regenerated using a district heating supply which in winter provides the heat source for the air handlers and radiators. The heat required to dry the desiccant thermal wheels in summer is essentially a by-product of electricity generation for the local grid, adding very little $CO_2$ to the atmosphere that would not already be produced for electricity. During the heating season, exhaust air is pulled through a heat-recovery unit which preheats incoming fresh cold air and supports the building's warm water systems. The preheated air is then distributed to the office spaces through local risers

located in the raised floor plenum (see Figure 2.6.10). Individually controlled radiators are provided along the building perimeter, sized for a −14 °C winter condition.

## Interface with the Central Building Management System

The building is designed to be completely naturally ventilated when external temperatures are between 5 °C and 25 °C. When the outdoor temperature drops below 5 °C or rises above 25 °C, mechanical ventilation is initiated by the building management system (BMS). The BMS controls the position of dampers located at the top and bottom of the thermal flue (see Figures 2.6.11 & 2.6.12) which regulates airflow from the building into the flue and within the flue according to the external weather conditions. The BMS gets input from an external weather station and from sensors evenly spread along the east and west façade, and inside the flue. These sensors monitor temperature, pressure, and air velocity. Depending on the input, the BMS indicates whether mechanical or natural ventilation is recommended to the users through the appearance of red or green lights on the window casement of each office module. In addition, a wall-mounted zone controller allows occupants to select individual zones within a floor to be either mechanically or naturally ventilated.

The transmission of heat and daylight into the interior is also controlled by the BMS, which is responsible for the operation of the moveable solar shading devices (vibrantly colored on the outward-facing side, adding character to the building's external aesthetic) located inside the flue of the west façade and venetian blinds located within the cavity of the east façade. In terms of controlling heat

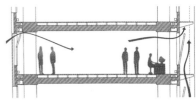

**(i) Open office**

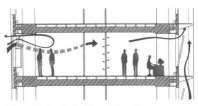

**(ii) Double-banked with cellular offices**

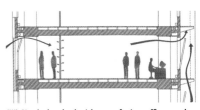

**(iii) Single-banked with west-facing offices and a corridor along the east**

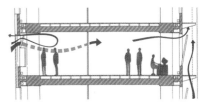

**(iv-a) Combi-east cellular offices along east façade with open-plan along the west façade**

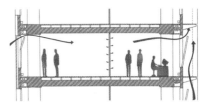

**(iv-b) Combi-west cellular offices along west façade with open-plan along the east façade**

▲ Figure 2.6.8: Cross ventilation section diagrams for the various floor plan configurations.
© Sauerbruch Hutton

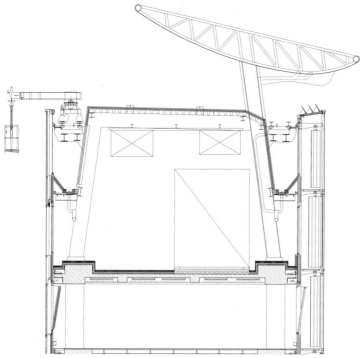

▲ Figure 2.6.9: Detailed section showing the Wing Roof above the thermal flue on the west façade. © Sauerbruch Hutton

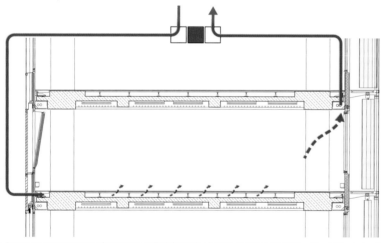

▲ Figure 2.6.10: In winter, exhaust air is pulled through a heat-recovery unit which preheats incoming fresh cold air which is then distributed to the office spaces through the raised floor plenum. © Sauerbruch Hutton

transmission and reducing solar heat gain, the blinds are automatically activated to avoid overheating in the morning or afternoon. Daylight transmission is encouraged through the use of photocells (light sensors) inside the façade which automatically switch off the row of electric lights adjacent to the windows when the blinds are open and daylight illumination is sufficient. Occupants can override the BMS's control of the blinds and the automated daylight-linked switching.

## Other Sustainable Design Elements

The double-skin façade acts as a thermal and acoustic buffer, protecting the building against heat loss, high wind speeds, and traffic noise. Perforated aluminum shutters are incorporated within the intermediate space of the western double-skin façade. The heat accumulated from the shutters and the sun is carried out of the building through the thermal flue of the west façade, minimizing solar heat gain and

reducing thermal load. The thermal performance of the building is affected by its exposed concrete structure, which moderates internal temperatures through radiant and convective heating and cooling. During the summer, the mass is cooled by night ventilation as the air sweeps along the exposed concrete ceiling. Night cooling draws out the excess heat generated during the day, providing greater comfort levels for users the following morning. In winter, the slabs help store the heat.

## Analysis – Performance Data

It is estimated that the building is naturally ventilated for around 70 percent of the year, when the outdoor temperature is between 5 °C and 25 °C. Extensive analysis using wind tunnel tests and in-house software (ROOM, FACMAX, FAVON, and VENT) was used during the design phase to size the necessary passive ventilation elements, including the optimal thermal flue depth, its height to highest discharge point, the size and control of the flue's top and base dampers, the size of the window opening (or more precisely, the profile of the opening surface), etc. With the dynamic thermal analysis program ROOM, it was concluded that between four and six air changes would give occupants an acceptable range of temperature, typically between 20 °C and 27 °C. Two programs, FACMAX and FAVON, were used to analyze the thermal flue, such as calculating the buoyancy available and the optimum width and height of the flue. The effects were modeled for a combination of weather data ranging from sunny to overcast days. Determining the height of the thermal flue above the highest occupied floor was important as it needed to be sufficient to avoid warm air re-entering the building at the top floors. It was concluded that a one-meter-deep thermal flue extending seven meters above the uppermost floor was optimum.

Through analysis, it was evident there were adverse conditions that would cause backflow into the floors toward the bottom of the building. Similarly, excessive air changes could occur at lower levels during very sunny weather. Both conditions were avoided by maintaining a 10 °C temperature difference between the top and bottom dampers in the flue. When the temperature rises, the louvers admit more air into the flue and conversely when the temperature falls, they close and trap air in the flue. Thus, both wind and solar effects are accounted for simultaneously.

Finally, CFD analysis was undertaken to ease concerns that a fire located on a lower floor would emit smoke back into the building at a higher level through the thermal flue. The conclusion was that energy within the smoke was beneficial in producing sufficient thermal buoyancy, which in turn helped break down the smoke within the flue (Cziesielski 2003).

### Analysis – Strengths

▸ The narrow plan facilitates the flow of air across the space and enhances the effectiveness of natural ventilation.

▸ The thermal flue ensures the entry of fresh air from the east even if the prevailing winds are traveling at a low speed or from a different direction. In such a case, buoyancy-driven ventilation in the thermal flue will assist wind driven natural ventilation. The thermal flue enables the control of natural ventilation (with air exchange rates comparable to mechanical air exchange rates) without the negative effects of cross currents.

▲ Figure 2.6.11: Dampers located at the top of the thermal flue. © Arup

▲ Figure 2.6.12: Dampers located at the bottom of the thermal flue. © Annette Kisling

▲ Figure 2.6.13: View of the Wing Roof which sits above the thermal flue on the west façade.
© Janice Ninan, J-Space Studio

**When the temperature rises, the louvers admit more air into the flue and conversely when the temperature falls, they close and trap air in the flue. Thus, both wind and solar effects are accounted for simultaneously.**

▸ As the sun warms the air within the intermediate space of the flue, the shutters heat up and stack effect is improved. As a result of the temperature increase within the flue, cool fresh air will be drawn in at an ever-faster rate, enhancing natural ventilation. The heat absorbed by the shutters can then be removed by convection as the air is exhausted at the top (Arons & Glicksman 2008). The greater the incident of solar radiation, the greater the rate of upward air movement in the thermal flue of the west façade. Ventilation rates and comfort levels are thus maintained even on hot days.

▸ The Wing Roof (see Figure 2.6.13) prevents the risk of down currents from occurring regardless of the prevailing wind direction. When the wind is blowing from the east or west direction, the aerodynamic shape of the Wing Roof accelerates the wind passing over the thermal flue. Conversely, if the wind is blowing from the northerly or southerly directions, a series of fins suspended from the roof's wing cause the wind to eddy. Both scenarios prevent the risk of a positive pressure build up and the presence of down currents over the outlet of the thermal flue (Hodder 2001).

▸ Night ventilation, when used in conjunction with the exposed building mass, is an effective cooling strategy in the summer. Furthermore, the thermal flue carries away excessive heat generated by the sun and the offices' internal heat gains. These factors affect the thermal performance of the building, minimizing its cooling loads and the risk of overheating. As a result, the airflow needed for comfort ventilation is reduced, ensuring the effectiveness of natural ventilation even when wind speeds are low or when the wind is not blowing from the east (where the air intake openings are located), as the ventilation concept is more or less independent of external conditions.

## Analysis – Considerations

The following areas could be causes for concern if adopting similar strategies in other buildings and should therefore be considered:

▸ When natural ventilation has to pass through partitions and corridors (such as the double-banked office plan), pressure issues may cause insufficient ventilation rates, problems such as doors slamming shut, or large drafts. Careful planning must be considered to ensure controlled air volume flow rates through these spaces to avoid such problems.

▸ The Venturi effect generated by the Wing Roof is more efficient if the prevailing wind direction is from the east or west. Therefore, sufficient uplift force in the thermal flue is only provided from easterly or westerly wind.

▸ The ventilation strategy might not function effectively if there is no sun to induce the thermal flue on the west façade (especially if wind speeds are low under overcast sky conditions). In this case, design considerations should be taken for the opening sizes that are needed to achieve an inward flow of fresh air which is induced by buoyancy alone due to differences between internal and external temperatures.

▸ If the top and bottom dampers of the thermal flue are not properly controlled and regulated, the double-skin façade might face the risk of overheating in the summer.

▸ Existing and future developments, such as the erection of a high-rise adjacent to the site might cast a shadow on the thermal flue façade or block the amount of fresh air entering the building, and thus have a negative impact on the performance of the thermal flue.

▸ Even though the narrow plan facilitates natural ventilation, it may limit the efficient use of the floor plate.

## Project Team

**Owner/Developer:** Gemeinnützige Siedlungs- u. Wohnungsbaugesellschaft mbH (GSW)
**Design Architect:** Sauerbruch Hutton
**Structural Engineer:** Arup
**MEP Engineer:** Arup
**Environmental Services:** Arup
**Main Contractor:** ARGE Züblin/Bilfinger Berger
**Other Consultants:** Emmer, Pfenninger + Partner (Façade Consultants)

**References & Further Reading**

**Books:**

▸ Cziesielski, E. (ed.) (2003) *Bauphysik-Kalender 2003*. Ernst & Sohn: Berlin, pp. 633–646.

▸ Sauerbruch, M. & Hutton, L. (eds.) (2000) *GSW Headquarters, Berlin. Sauerbruch Hutton Architects*. Lars Müller Publishers: Baden.

▸ Wigginton, M. & Harris, J. (2004) *Intelligent Skins*. Architectural Press: Oxford, pp. 49–54.

**Journal Articles:**

▸ Clemmetsen, N., Muller, W. & Trott, C. (2000) "GSW headquarters, Berlin," *Arup Journal*, vol. 35, no. 2, pp. 8–12.

▸ Hodder, S. (2001) "GSW Headquarters, Berlin; architects: Sauerbruch Hutton," *Architecture Today*, vol. 116, pp. 30–49.

▸ Pepchinski, M. (2002) "GSW Headquarters, Berlin," *A&U: Architecture and Urbanism*, vol. 384, no. 9, pp. 64–73.

▸ Russell, J. (2000) "GSW Headquarters, Berlin," *Architectural Record*, vol. 188, no. 6, pp. 156–164.

▸ Sauerbruch, M. & Hutton, L. (2001) "GSW Headquarters, Berlin, Germany; Sauerbruch Hutton Architects," *UME*, vol. 13, pp. 24–37.

## Project Data:

**Year of Completion**
- ▸ 2002

**Height**
- ▸ 163 meters

**Stories**
- ▸ 42

**Gross Area of Tower**
- ▸ 65,323 square meters

**Building Function**
- ▸ Office

**Structural Material**
- ▸ Composite

**Plan Depth**
- ▸ 12 meters (from central void)

**Location of Plant Floors:**
- ▸ 20

## Ventilation Overview:

**Ventilation Type**
- ▸ Mixed-Mode: Zoned / Complementary-Changeover

**Natural Ventilation Strategy**
- ▸ Cross and Stack Ventilation (connected internal spaces)

**Design Strategies**
- ▸ Double-skin façades
- ▸ Full-height central atrium divided into 9-story sky gardens
- ▸ "Wing Wall" extensions
- ▸ Aerodynamic external form

**Double-Skin Façade Cavity:**
- ▸ Depth: 1.7 meters (south façade), 1.2 meters (north façade)
- ▸ Horizontal Continuity: Fully Continuous (along entire length of façade)
- ▸ Vertical Continuity: Approximately 32 meters (height of sky gardens)

**Approximate Percentage of Year Natural Ventilation can be Utilized:**
- ▸ Unpublished

**Percentage of Annual Energy Savings for Heating and Cooling:**
- ▸ 79% compared to a fully air-conditioned German office building (measured)

**Typical Annual Energy Consumption (Heating/Cooling):**
- ▸ 75 kWh/m² (measured)

# Post Tower Bonn, Germany

## Climate

The climate and temperature of Bonn are often influenced by the nearby Rhine Valley and strong westerly maritime winds which blow in from the North Sea. In general, the weather is characterized by four distinct seasons and cloudy skies are usually the norm. Winter temperatures average around 3 °C, climbing to 10 °C by springtime. Snowy weather is light with rainy days being more likely. Summer temperatures regularly top 20 °C, rising to more than 25 °C at times, with long spells of sunshine. However, rainy weather occurs often with quick, unexpected showers (see Figure 2.7.1).

## Background

The Post Tower in Bonn, Germany, was completed in 2002 (see Figure 2.7.2). The Tower's form consists of two offset elliptical segments separated by a 7.2-meter-wide atrium that faces west toward the City of Bonn, and east towards the Rhine River. This full-building-height atrium is segmented

into four sky gardens; three are nine stories high and the top is 11 stories high. In each elliptical segment of the building, cellular offices hug the perimeter, with conference rooms and core functions located toward the center of the ellipse (see plan and section, Figures 2.7.3 & 2.7.4). The two elliptical halves of the tower are connected at every level across the atrium by steel and glass bridges that access the elevator lobbies. The two main façades of the office segments face north and south, respectively, while the façades of the sky gardens have east and west orientations.

## Natural Ventilation Strategy

From the onset of planning, there was a strong desire to give all office staff direct access to the outside air, including individual control of this access. Other key client criteria included transparent, floor-to-ceiling glazing and natural light in all office spaces. The natural ventilation strategy of the building is based on a double-skin façade system which supplies air to the

## Climatic Data:[1]

**Location**
- Bonn, Germany

**Geographic Position**
- Latitude 50° 43′N, Longitude 7° 5′E

**Climate Classification**
- Temperate

**Prevailing Wind Direction**
- West-northwest

**Average Wind Speed**
- 2.4 meters per second

**Mean Annual Temperature**
- 10 °C

**Average Daytime Temperature during the Hottest Months (June, July, August)**
- 17 °C

**Average Daytime Temperature during the Coldest Months (December, January, February)**
- 3 °C

**Day/Night Temperature Difference During the Hottest Months**
- 11 °C

**Mean Annual Precipitation**
- 796 millimeters

**Average Relative Humidity**
- 72% (hottest months); 83% (coldest months)

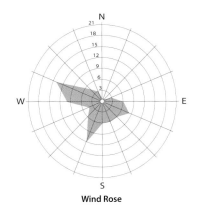

Wind Rose

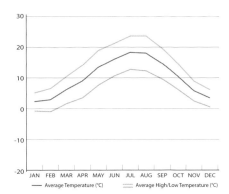

Average Annual Temperature Profile (°C)

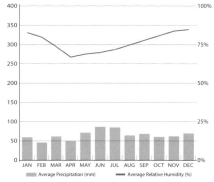

Average Relative Humidity (%) and Average Annual Rainfall

▲ Figure 2.7.1: Climate profiles for Bonn, Germany.[1]
◀ Figure 2.7.2: Overall view. © Murphy/Jahn Architects

[1] The climatic data listed for Bonn was derived from the World Meteorological Organization (WMO) and Deutscher Wetterdienst (German Weather Service).

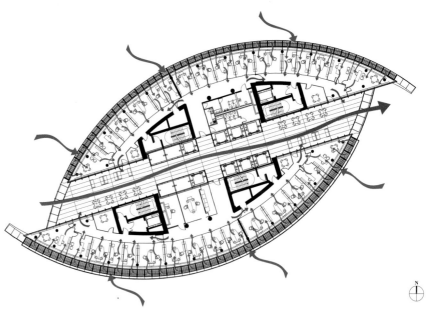

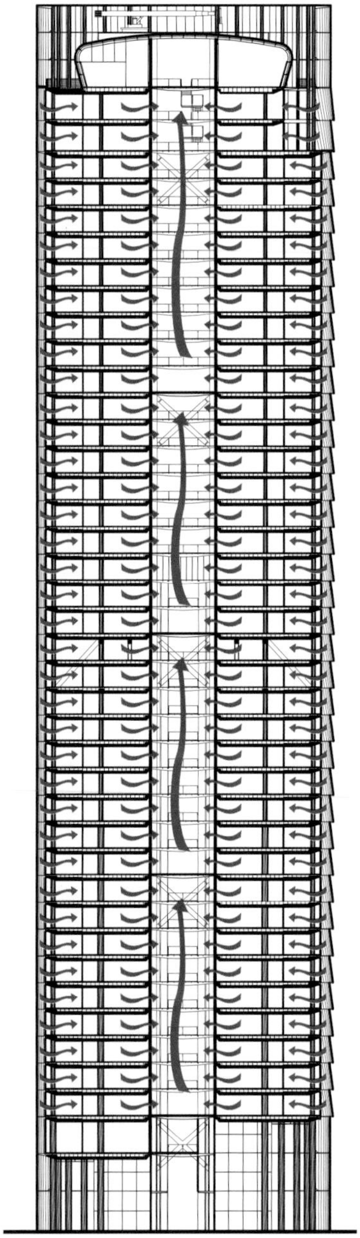

## Plan

The offices are ventilated through pivoting horizontal flaps in the outer skin of the double-skin façade. Cross-ventilation draws the air through the offices into the corridor where it is then exhausted into the nine-story atrium. Vents located at the top and bottom of the atrium façade contribute to stack effect in the atrium, aiding in exhausting of air from the building.

## Section

Each office floor is cross-ventilated by drawing fresh air in from the double-skin façade and exhausting it into the sky gardens (through vents in the raised floor system since the sky gardens are separated from the corridors with glazing). Additional vents at the bottom of each sky garden façade add cool air to aid stack effect within the nine-story space and vents at the top of each sky garden façade exhaust the air from the building.

▲ Figure 2.7.3: Typical office floor plan.
◀ Figure 2.7.4: Building section. (Base drawings © Murphy/Jahn Architects)

▲ Figure 2.7.5: Interior view of façade from inside sky garden, showing "wind needles" located at each floor level supporting the façade. © Murphy/Jahn Architects

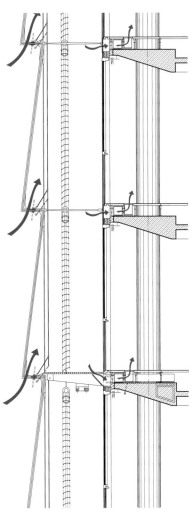

▲ Figure 2.7.6: Detailed section through the south façade showing operable flaps to let air into the cavity. © Murphy/Jahn Architects

offices, and a central atrium sky garden which exhausts it.

The double-skin façade varies in both width and detail on the north and south façades; however, both façades consist of an outer single-glazed skin, full-floor operable sunshades in the cavity, and an inner skin of floor-to-ceiling insulated double-glazing. The laminated glass panels of the outer skin are hung from extruded stainless steel sections supported at each sky garden level by steel brackets cantilevering from the concrete slab. This efficient, minimal structure resolves the vertical gravity load of the façade. Horizontal wind loads are transferred back to the structure with minimal "wind needles" located at each floor level (see Figure 2.7.5). The double-skin façade extends both vertically and horizontally along the north and south façades, with a single-skin façade enclosing the sky gardens between. The cavity is segmented vertically, coinciding with the sky gardens every nine floors (11 floors for the uppermost sky garden).

The outer skin acts as a protective layer mitigating high wind speeds at upper elevations and enabling operable windows of the inner skin to naturally ventilate the office spaces. By shielding the wind, it achieves wind pressures / airflow rates similar to a low-rise office building. Through stack effect within the nine-story-high double-skin façade cavity, operable flaps (see Figure 2.7.6) at the lower floors draw in cooler air while the upper floors exhaust hot air. On hot summer days, the operable flaps are opened to their maximum angle to ensure sufficient airflow within the cavity. During the winter, these openings are closed to a minimum so the façade cavity acts as a thermal buffer, reducing heat loss. Aside from temperature, the flaps can be adjusted to evenly distribute wind pressure along the inner façade, allowing everyone to open their windows even with high external wind speeds.

The outer shell on the north and south sides of the building is configured differently to address the two exposures.

The natural ventilation strategy of the building is based on a double-skin façade system which supplies air to the offices, and a central atrium sky garden which exhausts it.

▲ Figure 2.7.7: Exterior detail view of the south façade, which has sloped glass panes.
© Murphy/Jahn Architects

▲ Figure 2.7.8: Exterior detail view of the north façade, which has a smooth surface.
© Murphy/Jahn Architects

▲ Figure 2.7.9: View into the façade cavity (1.7 meters deep) of the south façade.
© Murphy/Jahn Architects

▲ Figure 2.7.10: View into the façade cavity (1.2 meters deep) of the north façade.
© Murphy/Jahn Architects

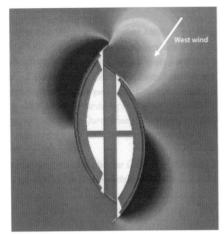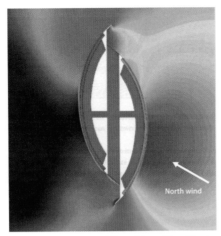

▲ Figure 2.7.11: CFD models showing the effect of the Wing Walls to create pressure differentials at either side of the building to induce cross-ventilation of the central sky garden spaces. © Transsolar

**CFD studies have shown that the Wing Walls help create pressure differentials at each sky garden façade to induce natural cross-ventilation through these spaces in the center of the building.**

The south façade features sloped glass panes allowing air intake and extract through flaps at the bottom of each panel (see Figure 2.7.7). Conversely, the north façade is a smooth, vertical plane with flaps at the bottom of every façade panel (see Figure 2.7.8). Since the south façade receives direct sunlight, more air must flow through its cavity than in the north façade to avoid overheating. The differing depth of the cavities (the south façade is 1.7 meters and the north is 1.2 meters) allows for different required airflow to vent excess heat (see Figures 2.7.9 & 2.7.10).

The outer skin of the double-façade extends beyond the building to create "Wing Walls" (at the east end of the north façade and the west end of the south) that improve the aerodynamic nature of the building and contribute to the natural ventilation strategy. CFD studies have shown that these Wing Walls help create pressure differentials across opposite sky garden façades to induce natural cross-ventilation through the atria (see Figure 2.7.11).

The inner façade consists of an argon-filled, double-glazed aluminum curtain wall system. Motorized, side-hung operable windows are located in every other façade module, which can be either manually operated by a local control panel or centrally controlled. After normal working hours the windows can be centrally opened to provide night flushing with cool air.

The overall natural ventilation strategy relies on both cross and stack ventilation. Outside air enters the building through the double-skin façade, flows from the offices into the corridors and is exhausted through the sky gardens. The flow of air from the office spaces into the corridors is enabled through open joints in the office partition walls, and "loose" sliding doors, with a 20 mm gap at both top and bottom of the door (see Figure 2.7.12).

The adjoining corridor thus acts as a horizontal exhaust air duct that vents the exhaust air into the sky garden. Since the corridor is separated from the sky garden by glazing (see Figure 2.7.13) exhaust air flows through a vent and fire damper in the ceiling slab to a fan located in the raised floor system above, which aids in exhausting the air into the atrium. Since air exhausted from the office corridors is heated due to internal gains, it tempers the air in the sky garden. Through stack effect, exhaust air is vented through operable windows located high in the façade of each sky garden, with fresh air added through low-level vents to assist the natural airflow (see Figure 2.7.14). The walkways that connect the north offices to the south are open bridges and therefore allow cross ventilation in the sky gardens (see Figure 2.7.15).

**Mixed-Mode Strategy**

The Post Tower is designed as a Zoned/ Complementary-Changeover building. While interior conference and meeting room spaces are always conditioned mechanically, the exterior cellular offices benefit from either natural or mechanical ventilation. During the extremes of summer or winter, thermal conditioning of the offices is augmented by both perimeter fan coil units and radiant ceilings. Perimeter fan coil units are located below the floor, adjacent to every other façade module and can be individually controlled in each office. The fan coils draw in outside air from the double-skin façade and then heat or cool the air according

▲ Figure 2.7.12: Open joints in the office partition walls, and "loose" sliding doors allow the flow of air from the office spaces into the corridors. © Murphy/Jahn Architects

▲ Figure 2.7.13: View of the end of a corridor which is separated from the sky garden atrium by glazing, with a fully enclosed, mechanically ventilated meeting room to the right. © Murphy/Jahn Architects

▲ Figure 2.7.14: Detail façade view showing the operable windows at the top and bottom of the sky gardens and the Wing Wall façade extensions. © Murphy/Jahn Architects

to demand. During cold winter days, air is warmed in the double-skin façade (through solar heat gain) before it enters the fan coil units. Additional conditioning is provided through radiant, exposed concrete soffits with embedded piping that circulates cool water (18 °C) during summer, and warm water (28 °C) during winter. The sky gardens rely fully on natural ventilation and are not mechanically ventilated.

Throughout most of the year, the only primary energy used for ventilation and general thermal conditioning is electricity to drive the water pumps for the radiant slabs, and the perimeter fan coils. During the cooling season, the fan coil units and the radiant slab system both utilize a ground water exchange with cool water from the Rhine River as an energy source. This eliminates the need for additional cooling for chillers. For heating, the energy source is district heating provided by the City of Bonn, sourced from waste heat produced by electricity production.

## Interface with the Central Building Management System

A computerized building management system (BMS) responds to local sensors controlling the operation of the outer façade flaps (according to external weather conditions: temperature, rain, noise, and wind speed) and the motorized, operable windows of the inner façade to adjust natural ventilation inside the office spaces. The control of vents on the outer façade maintains a low pressure difference to ensure proper ventilation rates. The BMS ensures the tower is operating under the designed temperatures range of 22 °C and 26 °C in the office spaces and 18 °C and 28 °C in the sky gardens. Additionally, the BMS controls the radiant concrete slabs, the sunshades located in the double-skin cavity, the dimming of the artificial lighting, and the vents located in the exterior façade of the sky garden.

While the building has a high level of centralized control, office occupants can override the BMS to ensure their individual comfort through a control panel located next to every door. This allows users to operate the blinds, control lighting levels, operate windows, and regulate internal temperatures. If the occupants decide to open the inner façade windows, the BMS will automatically switch off the perimeter fan coil units.

## Other Sustainable Design Elements

Highly reflective and adjustable blinds are integrated within the cavity of the double-skin façades on both sides of the building. These shading devices protect the office spaces against direct solar heat gain and help reduce glare inside the offices. Since these blinds are perforated, they offer good views to the surrounding landscape even when they are in the fully closed position. The underside of the concrete slabs are left exposed in the office space so that their high thermal capacity can be used to absorb and store a large amount of heat energy during the day. Night flushing cools the exposed slabs after working hours, allowing them to slowly absorb the heat generated in the office interior during the day. This regulates the internal temperatures and reduces the cooling loads during the summer. Thermal conditioning is further utilized by heating or cooling the water in the radiant ceiling pipes, allowing the whole ceiling to temper the surrounding spaces. The office spaces can also operate in "stand-by mode" when rooms are not occupied due to holiday or sick leave, which further increases energy savings.

## Analysis – Performance

A study was conducted in 2003 evaluating the efficacy of the design. While the energy levels did not achieve the aggressive benchmarks, they remained very impressive. The building was designed to utilize 65 kWh/m² for ventilation, heating, and lighting. The benchmark was an 83 percent reduction in energy consumption compared to a typical air-conditioned building and a 63 percent reduction compared to a "good practice office building." The highly efficient radiant slabs, the

exterior sunshades, and the use of natural ventilation and decentralized mechanical conditioning actually consumed 75 kWh/m² during the year 2003. This is a 79 percent reduction compared to a typical air-conditioned office building (see Figure 2.7.16).

## Analysis – Strengths

▸ The building's configuration and layout facilitate natural ventilation, with all offices located along the building's periphery within six

▲ Figure 2.7.15: Interior view of one of the sky gardens. The walkways that connect the north and south offices are open bridges that allow for cross ventilation of the space. © Murphy/Jahn Architects

**If outside air with a high relative humidity is introduced to an interior space with chilled radiant slabs, there is the potential for condensation to form on the radiant surface.**

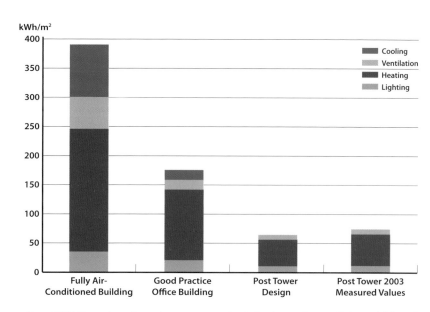

▲ Figure 2.7.16: Comparison of annual energy consumption for heating, cooling, ventilating, and lighting for Post Tower. (Source: Transsolar)

meters from operable windows. The sky gardens play an important role in the ventilation strategy, acting as extract chimneys that use stack effect to exhaust air out of the building. The sky gardens also provide communal gathering spaces and clearly organize the building's horizontal and vertical circulation.

▸ The double-skin façade provides an effective shield against high wind speed, noise, and rain. It acts as a thermal buffer in winter, and protects the exterior sunshades from strong winds, allowing them to be used effectively in all weather conditions. The extensions of the outer skin of the façades beyond the plan of the building creates Wing Walls that improve the aerodynamic nature of the building and strengthen cross-ventilation across the atrium sky gardens.

▸ The double-skin façade provides fresh air at every second façade module, with the sky gardens

acting as centralized extract chimneys. Combined, these two features eliminate the need for ventilation shafts in the core and major horizontal distribution ducts on every floor (since the ducts are provided by the corridors). As a result, the space needed for the main MEP plant has been cut in half and limited to just one floor.

▸ The horizontal segmentation of the building and façade into four nine-story zones prevents the development of extreme stack flows that might cause drafts and high pressures otherwise (Wells 2005).

▸ The exterior sunshades prevent the inner skin from solar heat gain, enabling relatively cooler air to flow into the interior.

▸ A decentralized fan coil unit coinciding with the radiant cooling system replace a traditional, centralized, ventilation-based air-conditioning system. This results in

significant energy, cost, and space savings. Further saving are generated due to the fan coils only being used when the room is in use.

▸ All of the building's cooling needs are supplied by a heat exchange system that draws water from the neighboring Rhine River. This allows the building to dispense with the typical mechanical cooling towers and their associated central plant.

▸ The BMS provides a high degree of centralized control, allowing the building to easily adjust to changing weather conditions. Individual control of the office spaces has also been prioritized by allowing the occupants to override the BMS.

## Analysis – Considerations

The following areas could be causes for concern if adopting similar strategies in other buildings and should therefore be considered:

- Due to stack effect in the intermediate space of the double-skin façade, the upper part of the nine-story-high façade section may experience an over-pressure that drives air out of the openings. On the other hand, the under-pressured zone at the bottom of the façade cavity may pull relatively colder air inwards. As a result, the lower levels of the nine-story-high sections may receive more fresh air than the top levels. Furthermore, as the air in the top part of the façade cavity is warmer than that at the bottom, the top stories could possibly overheat in summer (this is controlled in Post Tower by the fresh air being precooled by the fan coil units).

- When the wind meets an obstacle such as a tall building, some of it will be deflected upwards, but most of it will spiral downwards creating a so-called "standing vortex" or turbulence near the ground. This occurrence might impede air from flowing into the bottom part of the south façade since the horizontal openings of the sloping-shingles would be in the opposite direction of the flowing wind.

- The significant depth of the double-skin façade (1.2–1.7 meters) could make the solution commercially unviable in other buildings, especially those on constrained sites with restricted floor plates. In addition, the loss of usable floor space through the sky gardens could be deemed financially restricting by some developers.

- Relative humidity should be carefully considered when using radiant surface cooling such as activated structural slabs in combination with natural ventilation. If outside air with a high relative humidity is introduced to an interior space with chilled radiant slabs, there is the potential for condensation to form on the radiant surface. Post Tower was not subject to this concern because of the low relative humidity during the summer months.

- The design of a double-skin façade requires the commitment of a dedicated, long-term owner committed to the higher initial costs. In the case of Post Tower, the client realized the value of an investment in a headquarters that would minimize long-term energy consumption, while providing high-quality office space with a high level of individual control.

## Project Team

**Owner/Developer:** Deutsche Post AG
**Design Architect:** Murphy/Jahn Architects
**Associate Architect:** Heinle Wischer + Partner
**Structural Engineer:** Werner Sobek Group
**Energy Consultant:** Transsolar
**MEP Engineer:** Brandi Consult, GmbH
**Main Contractor:** Hochtief
**Other Consultants:** Arge Permasteelisa; Gartner (Façade Contractor); I.F.I. Aachen (Wind Engineering)

### References & Further Reading

**Books:**

- Blaser, W. (2003) *Post Tower: Helmut Jahn, Werner Sobek, Matthias Schuler.* Birkhäuser: Basel.

- Castillo, P. (2001) "Two towers," in Beedle, L. (ed.) *Cities in the Third Millennium – Proceedings of the CTBUH 6th World Congress.* Spon Press: London, pp. 413–416.

- Eisele, J. & Kloft, E. (2003) *High-rise Manual: Typology and Design, Construction and Technology.* Birkhauser: Basel, pp. 186–188.

- Wells, M. (2005) *Skyscrapers: Structure and Design.* Laurence King Publishing Ltd.: London, pp. 86–91.

**Journal Articles:**

- Buccino, G. (2004) "Deutsche Post, Bonn," *Industria delle Costruzioni,* vol. 38, no. 376, pp. 64–71.

- Dassler, F.H., Sobek, W., Reuss, S. & Schuler, M. (2003) "Post Tower in Bonn: state of the art," *XIA Intelligente Architetektur,* vol. 41, pp. 22–37.

## Project Data:

**Year of Completion**
- ▸ 2004

**Height**
- ▸ 180 meters

**Stories**
- ▸ 42

**Gross Area of Tower**
- ▸ 64,470 square meters

**Building Function**
- ▸ Office

**Structural Material**
- ▸ Steel

**Plan Depth**
- ▸ 6.4–13.1 meters (from central core)

**Location of Plant Floors:**
- ▸ 35


## Ventilation Overview:

**Ventilation Type**
- ▸ Mixed-Mode:
  Complementary-Concurrent

**Natural Ventilation Strategy**
- ▸ Cross and Stack Ventilation
  (connected internal spaces)

**Design Strategies**
- ▸ Double-skin façades
- ▸ Stepping atria which tempers air
  before being distributed to offices

**Double-Skin Façade Cavity:**
- ▸ Depth: 1–1.4 meters
- ▸ Horizontal Continuity: Varies (between
  diagonal structural frame members)
- ▸ Vertical Continuity: 4.15 meters
  (floor-to-floor)

**Approximate Percentage of Year Natural
Ventilation can be Utilized:**
- ▸ 40% (as originally designed)

**Percentage of Annual Energy Savings for
Heating and Cooling:**
- ▸ Unpublished

**Typical Annual Energy Consumption
(Heating/Cooling):**
- ▸ Unpublished

# 30 St. Mary Axe London, UK

## Climate

Summers in London can often get warm and humid, with an average high temperature around 22 °C. Although there is much less rain in the summer, the weather easily changes and it is not unusual for a sunny day to quickly turn rainy. Winters can be long and damp with humidity levels hovering around 80 percent. Snow is infrequent, but there are often cold, frosty days (see Figure 2.8.1).

## Background

30 St. Mary Axe is an office tower initially designed for the reinsurance company, Swiss Re. There was a desire to create an architecturally significant and iconic building, which was successfully achieved through the building's strong and unusual cylindrical form (see Figure 2.8.2). The form also contributes to the environmental strategy for the building, as well as responds to the tight site constraints at ground level. As it is located in the heart of London's financial center, the building is surrounded by historic buildings on a narrow street of medieval dimensions. Within its rectangular site, the tapering circular profile of the building allows an increase in the amount of daylight penetration at plaza level. The size of the floor plates vary, with the first floor having a 50 meter diameter, which widens to 57 meters on level 17, and then diminishes toward the apex (see plan and section, Figures 2.8.3 & 2.8.4).

Triangular atria (designed to play a key role in the natural ventilation strategy) cut into the circular floor plan, creating six rectangular-shaped office spaces (referred to as "fingers") that radiate from a central core containing elevators, stairs, toilets, and service risers. The distance between the core and building perimeter ranges from 6.4 to 13.1 meters, depending on the floor plate size in the tapering form. Each floor plate rotates five degrees with respect to the one below, forming stepped atria which spiral up the building, segmented every six floors to create "office villages."

## Climatic Data:[1]

**Location**
- London, UK

**Geographic Position**
- Latitude 51° 32′ N, Longitude 0° 5′ W

**Climate Classification**
- Temperate

**Prevailing Wind Direction**
- Southwest

**Average Wind Speed**
- 3.6 meters per second

**Mean Annual Temperature**
- 11 °C

**Average Daytime Temperature during the Hottest Months (June, July, August)**
- 22 °C

**Average Daytime Temperature during the Coldest Months (December, January, February)**
- 8 °C

**Day/Night Temperature Difference During the Hottest Months**
- 9 °C

**Mean Annual Precipitation**
- 611 millimeters

**Average Relative Humidity**
- 66% (hottest months); 81% (coldest months)

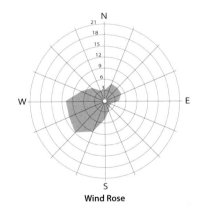

**Wind Rose**

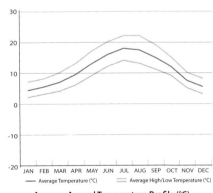

**Average Annual Temperature Profile (°C)**

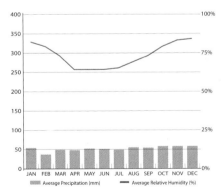

**Average Relative Humidity (%) and Average Annual Rainfall**

▲ Figure 2.8.1: Climate profiles for London, UK.[1]
◀ Figure 2.8.2: Aerial view from the west. © Nigel Young / Foster + Partners

[1] The climatic data listed for London was derived from the World Meteorological Organization (WMO) and the MET Office (United Kingdom Weather Service).

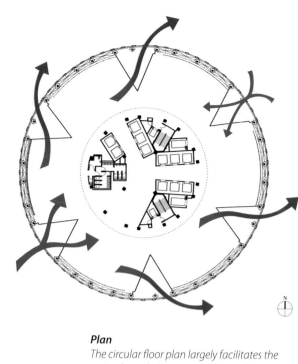

## Plan

The circular floor plan largely facilitates the natural ventilation strategy, increasing pressure differentials between the windward and leeward sides, inducing cross-ventilation. Fresh air enters through the windward-facing atria and exhausts out the leeward-facing atria. Air also enters each six-story atria and rises through stack effect. Each office "finger" thus receives fresh air from the adjacent atrium.

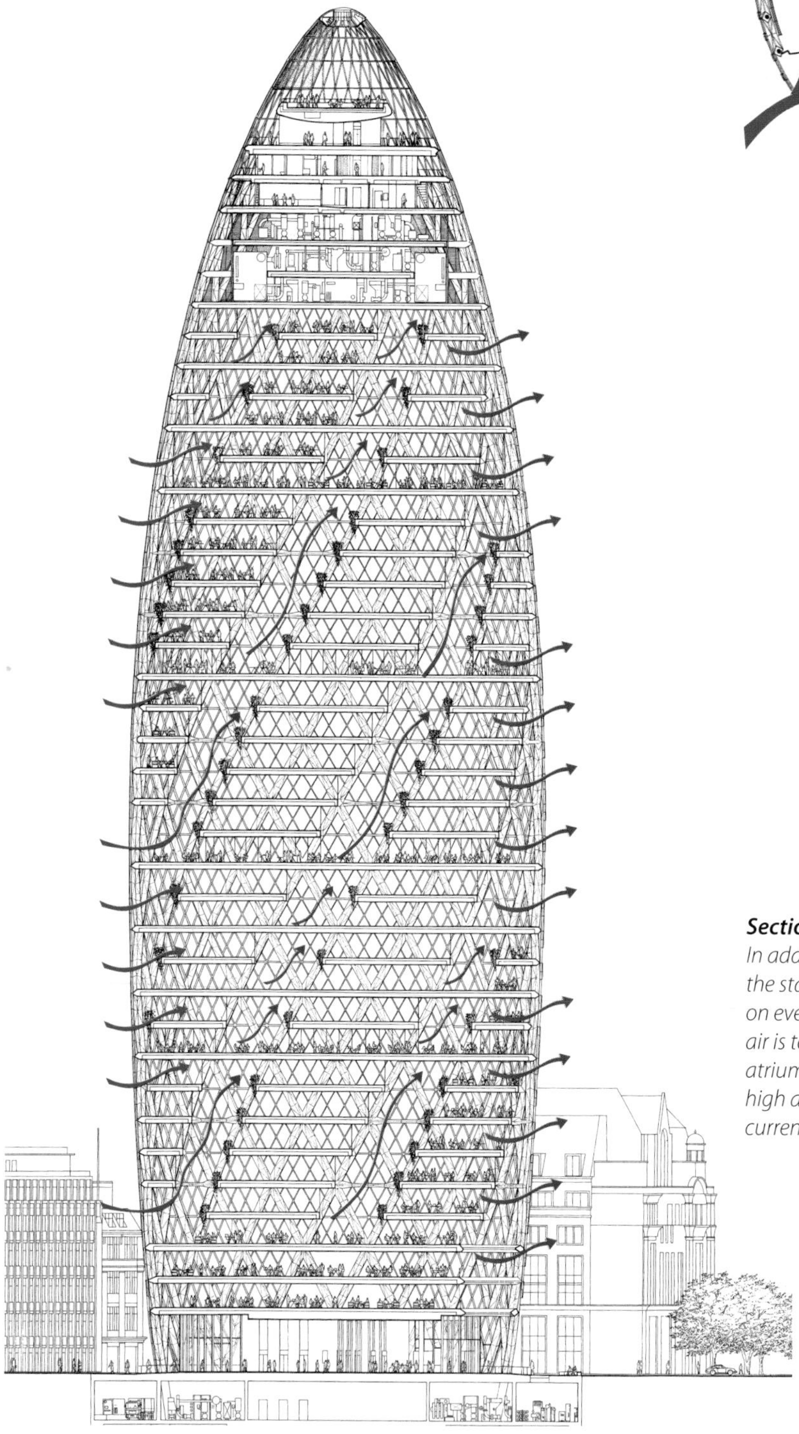

## Section

In addition to cross-ventilation, stack effect is exploited through the stacked, spiraling atria. Triangular-shaped windows located on every other floor induce fresh air into the atrium where the air is tempered before being distributed to the offices. As each atrium spirals around the building, they take advantage of the high and low pressure zones of the façade created by wind currents flowing around the aerodynamic tower.

▲ Figure 2.8.3: Typical office floor plan.
◀ Figure 2.8.4: Building section. (Base drawings
© Foster + Partners)

▲ Figure 2.8.5: Façade view showing atrium windows in the open position. © Nigel Young / Foster + Partners

**The aerodynamic curved form of the building largely facilitates the natural ventilation strategy by increasing the wind velocity and pressure differentials between the windward and leeward sides.**

Acting as both a cladding and structural element, the envelope consists of a diagonally braced steel diagrid which wraps the tower, supports the floors at their perimeter and allows for a column-free internal space.

## Natural Ventilation Strategy

The aerodynamic curved form of the building largely facilitates the natural ventilation strategy. As wind flows around the tower it accelerates, increasing the pressure differentials between the windward and leeward sides for more effective cross-ventilation while reducing downdrafts and causing less turbulence at street level.

In addition to cross-ventilation, stack effect is also exploited through stacked atria, which induces and tempers fresh air before distributing it to the office spaces, and protects the office spaces from direct winds. The outer façade perimeter of each atrium features triangular-shaped motorized windows which are arranged in groups of four top-hung and four bottom-hung windows on every other floor (see Figures 2.8.5 & 2.8.6). As each atrium spirals around the building's perimeter,

some windows face the windward side while others face the leeward side of the tower, taking advantage of high and low pressure zones created on the façade (Gregory 2004). In this regard, both the geometry of the tower and the configuration of the atria exploit the pressure differentials around the perimeter of the building to enhance natural ventilation within it. It is also worth noting that the atria are intentionally rotated clockwise along with the prevailing southwest winds in order to optimize natural ventilation (Jenkins 2009).

The atria are segmented into "villages" of two or six floors (see Figure 2.8.7) where the base of each atrium "village" can be utilized as an informal meeting space. In addition to fire compartmentalization, these vertical segmentations prevent strong updrafts due to large pressure differentials caused by the stack height.

The office working areas are not ventilated through the exterior façade as the atria. They have operable windows in the internal façade enclosing the atria. In an open-plan layout, the windows are used to bring in fresh air from the windward atrium and exhaust air into the leeward atrium. Single-sided

ventilation was also tested and shown to be effective, where offices only had operable windows facing a single six-story atrium.

The exterior-facing side of the office spaces feature an active, ventilated façade which consists of a clear, low-E double-glazed layer on the outside and a single-glazed inner screen. The width of the double-skin cavity ranges between 1 and 1.4 meters and accommodates venetian solar-control blinds which reduce glare and solar heat gain. The double-skin cavity of the offices is closed to the atrium by the diagonal structure. Exhausted air from the office spaces is channeled through the cavity, which mitigates solar heat gain in the summer or warms the cavity during the winter.

## Mixed-Mode Strategy

Designed to be naturally ventilated without supplemental heating or cooling for approximately 40 percent of the year (Jenkins 2009), 30 St. Mary Axe is equipped for mechanical and natural ventilation to operate simultaneously as a Complementary-Concurrent system. However, when external conditions are

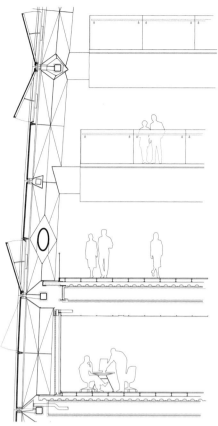

▲ Figure 2.8.6: Detailed section showing the alternating arrangement of bottom-hung and top-hung windows. © Foster + Partners

▲ Figure 2.8.7: Internal view of a six-story atrium – looking downwards onto the social meeting space at its base. © Nigel Young / Foster + Partners

**The outer façade perimeter of each atrium features triangular-shaped motorized windows which are arranged in groups of four top-hung and four bottom-hung windows on every other floor.**

too hot or cold, too windy or rainy, the building can be sealed and mechanical conditioning will be activated. The top three floors of the tower (housing private dining and meeting rooms) are isolated and entirely air-conditioned.

Although heavy plant equipment is located in the basement with six closed-circuit cooling towers located on level 35, a decentralized mechanical system serves each floor separately. Air-handling units (AHU) contained in ceiling voids allow for each office floor or each office "finger" (divided by the atria) to be individually controlled. As conditioned air is introduced and extracted on a floor-by-floor basis when needed, it allows the office environment to be easily adjusted and fine-tuned to the need of users, thus optimizing energy consumption (Powell 2006).

When operating under mechanical ventilation, external fresh air is drawn in through narrow slits between the

glazing panels at floor level (each office finger has an intake and extract slot). It should be noted that these slits are only used for the mechanical ventilation system and are not part of the natural ventilation system. The air is then conditioned through an AHU in the ceiling void before being introduced into the office space at ceiling level (see Figure 2.8.8). Exhaust air is divided with part returning to the AHU while the remainder is drawn through the raised floor and exhausted into the "ventilated façade" cavity. While being drawn up between the blinds and the inner glazing, the air in the cavity is tempered.

During summer, conditioned air vented into the cavity cools the glass and carries away absorbed heat from the blinds. During winter, the air vented into the façade cavity minimizes the chilling effect of the glass. Before warm air is extracted, it passes through the thermal wheels of a heat-recovery unit, maximizing the efficiency of the air-conditioning system.

## Interface with the Central Building Management System

30 St. Mary Axe is equipped with a building management system (BMS) that monitors external weather conditions and controls the operation of window openings. When the external temperature is between 5 °C and 28 °C, the building is naturally ventilated, maintaining internal temperatures between 20 °C and 24 °C. When interior temperatures rise above 24 °C, mechanical air-conditioning is activated while the windows remain open. If external temperatures exceed 28 °C, the BMS closes all windows and fully air-conditions the entire building. When external temperatures fall below 12 °C, the building supplements natural ventilation with active heating. Furthermore, if external temperatures fall below 5 °C, the building is entirely sealed and mechanically heated (Jenkins 2009). Outdoor relative humidity higher than 60 percent and wind speeds higher than 10 m/s also trigger the closure of windows and the use of artificial systems (Gonçalves 2010).

In addition to mechanically/naturally conditioning each office segment on an individual basis (office space "fingers" separated by atria), the BMS controls the operation of blinds in the façade cavity to one of four settings: closed, one-third open, two-thirds open, and horizontal. The operation of the blinds cuts out direct sun angles above 22 degrees and eliminates approximately 85 percent of solar heat gain, while admitting 50 percent of natural light.

## Other Sustainable Design Elements

Several strategies were used to mitigate solar heat gain without limiting daylighting, including the circular form, which has 25 percent less surface area than a rectilinear building of the same

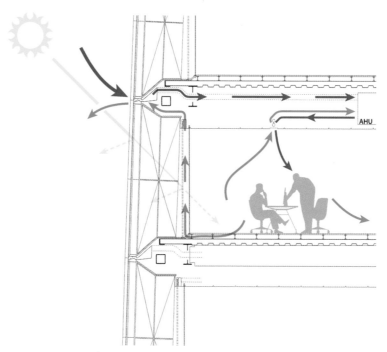

▲ Figure 2.8.8: Section illustrating the mechanical ventilation strategy in the summer. © Foster + Partners

volume. Blinds, which are located in the air cavity of the interior curtain wall, minimize solar heat gain in the office spaces. Chilled floors are used to maintain thermal comfort in the top-level restaurant. The atria also bring light deep into the office floors, reducing the dependency on artificial lighting (see Figure 2.8.9).

## Analysis – Performance

At the design stage an environmental assessment was carried out to determine the impact of the architectural design and building form on the natural ventilation strategy. The results indicated that the round and compact form would potentially achieve 20 percent in energy savings (Gonçalves 2010). Overall, the projected annual energy savings due to the environmental design were between 30 and 50 kWh/m², compared to the 250 kWh/m² of a similar fully air-conditioned office building in London, and the consequent reduction in $CO_2$ emissions was between 30 and 50 Kg $CO_2$/m² per year (Gonçalves 2010). The studies also revealed a 60 percent reduction in downdraft typical of a rectangular building with an equivalent volume.

Predictive studies also revealed the building's circular layout offered the possibility of longer periods of natural ventilation. Environmental assessment studies predicted the effectiveness of natural ventilation with regard to interior layout and concluded that 30 St. Mary Axe can be naturally ventilated for 41–48 percent annually, depending on external climatic conditions and open-plan versus cellular office layout. An internal temperature of 24 °C and an open-plan layout (both single-sided and double-sided ventilation) resulted in natural ventilation for 41 percent of the year, whereas a cellular layout with double-sided ventilation resulted in natural ventilation for 48 percent of the year (Gonçalves 2010). The outcome of the studies reveal that the building can be naturally ventilated for a longer period of the year with a cellular office configuration while maintaining the same thermal comfort standards.

Despite positive figures pertaining to the predicted energy performance, it is important to investigate post-occupancy performance, which is affected by factors such as operation and management routines, and occupant behavior. 30 St. Mary Axe was designed as an owner-occupied headquarters for

▲ Figure 2.8.9: Interior view of office balcony adjacent to one of the spiraling atria which allow natural light and air to the office floor plates. © Nigel Young / Foster + Partners

**Designed as an owner-occupied headquarters but ultimately converted into a multi-tenant building, the changes in tenancy patterns that were not accounted for in the design stages have compromised the effectiveness of the original natural ventilation strategy.**

Swiss Re where the natural ventilation strategy was a key design requirement from the client. In its realization, however, it was ultimately converted into a multi-tenant occupied building. This resulted in changes in tenancy patterns that were not accounted for in the design stages, including alterations of interior configurations and the addition of partitions and corridors due to privacy concerns with the atria. Such changes could restrict the predicted patterns of airflow within the building and potentially jeopardize the effectiveness of its natural ventilation strategy. In fact, most tenants have rejected the use of natural ventilation in favor of mechanical systems. Thus the performance analysis presented here is purely predictive as to how the building would have performed if it had been used as designed.

### Analysis – Strengths

▶ The aerodynamic form of the tower minimizes turbulence, improving pedestrian comfort at ground level and enhancing the possibilities of natural ventilation within the building. The more even distribution of pressures on the façade eliminates the need to over-design the envelope openings and optimizes its design.

▶ The cylindrical form requires 25 percent less external surface than a equivalent size rectilinear tower, thus minimizing heat gain and heat loss through the envelope.

▶ The spiraling of the atria generates pressure differentials that greatly assist the natural flow of air by combining cross-ventilation with the stack effect, enhancing the effectiveness of natural ventilation.

▶ Both the atria and the double-skin façade act as thermal buffer zones that provide additional insulation, regulate internal temperatures, and minimize energy consumption. The façade cavity of the offices is ventilated using extracted air from the offices, which is roughly the same temperature all year round, but relatively warmer than the outside temperature in winter and relatively cooler than the outside temperature in summer.

▶ The exterior-facing atria bring daylight and natural ventilation deep within the building while enhancing external views and visual communication between floors, and providing a space for social interaction.

▶ The triangular shape of the atria reconciles circular and rectilinear geometries allowing for rectangular offices in between. Thus, office spaces are easier to sub-divide for leasing purposes.

▶ The exterior "diagrid" provides structural support with column-free interior spaces and spiraling atria, natural ventilation is enhanced with minimal obstructions.

- The vertical segmentation of the atria allows for fire compartmentalized zones and prevents the development of extreme stacks/drafts typical in tall open spaces.

- The decentralization of the air-conditioning system allows the internal environment to be controlled on an individual basis where occupant comfort can be maximized and energy consumption can be optimized.

## Analysis – Considerations

The following areas could be causes for concern if adopting similar strategies in other buildings and should therefore be considered:

- The success of the natural ventilation strategy depends on how the occupants actually use the building. Generally, owner-occupiers tend to make more effective use of such energy-saving systems than tenants. The conversion of an owner-occupied into a multi-tenant building could compromise energy-saving goals, such as in the case of 30 St. Mary Axe.

- Alterations to internal layout and configuration (such as adding partitions and corridors) could reduce the designed effectiveness of natural ventilation by restricting internal airflow patterns.

- A building with the combination of a central core and spiraling atria is not spatially efficient by conventional economic standards. 30 St. Mary Axe resulted in only a 63 percent net-to-gross floor area (Spring 2008). When the building in operation does not use the actual natural ventilation systems as designed, the lost floor area serves less purpose.

- The building utilizes a closed double-façade system in the office spaces. In a temperate climate such as London, it could have been possible to open the façades to the outside to further enhance natural ventilation.

- Building occupants do not have access to operable windows, which could have enhanced occupants satisfaction and provided a feeling of having more control over their environment. This control can lead to an increased tolerance for a wide range of thermal comfort standards and allow an adaptation to temperature fluctuations in the building while the windows are open.

- The use of "office villages" such as the six-story spaces facing the atria in 30 St. Mary Axe, is good for visual communication, daylight and natural ventilation, but unless there are tenants large enough to occupy one entire village, these spaces could become compromised due to privacy concerns (as in 30 St. Mary Axe).

## Project Team

**Developer:** Swiss Re
**Design Architect:** Foster + Partners
**Structural Engineer:** Arup
**MEP Engineer:** Hilson Moran Partnership Ltd.
**Environmental Consultant:** BDSP Partnership Consulting Engineers
**Main Contractor:** Skanska AS
**Other Consultants:** RWDI (Wind Tunnel Testing)

**References & Further Reading**

**Books:**

- Gonçalves, J. (2010) *The Environmental Performance of Tall Buildings.* Earthscan Ltd.: London, pp. 251–257.

- Jenkins, D. (ed.) (2009) "Swiss Re headquarters," in *Norman Foster. Works 5.* Prestel: Munich, pp. 488–531.

- Lakkas, T. & Mumovic, D. (2009) "Sustainable cooling strategies," in Mumovic, D. & Santamouris, M. (eds.) *A Handbook of Sustainable Building Design & Engineering: An Integrated Approach to Energy, Health and Operational Performance.* Earthscan: London, pp. 287–290.

- Powell, K. (2006) *30 St Mary Axe: A Tower for London.* Merrell: London.

**Journal Articles:**

- Gonçalves, J. & Bodes, K. (2011) "The importance of real life data to support environmental claims for tall buildings," *CTBUH Journal,* vol. 2, pp. 24–29.

- Gregory, R. (2004) "Squaring the circle," *Architectural Review,* vol. 215, no. 1287, pp. 80–85.

- Spring, M. (2008) "Projects. Gherkin revisited; architects: Foster + Partners," *Building,* vol. 273, no. 8526(16), pp. 62–67.

## Project Data:

**Year of Completion**
- ▶ 2004

**Height**
- ▶ Tower 1: 126 meters
  Tower 2: 113 meters

**Stories**
- ▶ Tower 1: 33
  Tower 2: 28

**Gross Area of Towers**
- ▶ 74,148 square meters

**Building Function**
- ▶ Office

**Structural Material**
- ▶ Composite

**Plan Depth**
- ▶ 13.5 meters (between façades)

**Location of Plant Floors:**
- ▶ Semi-Decentralized/Every Floor

## Ventilation Overview:

**Ventilation Type**
- ▶ Mixed-Mode:
  Complementary-Concurrent

**Natural Ventilation Strategy**
- ▶ Wind-Driven Cross-Ventilation
- ▶ Stack Ventilation in Vertical Shaft in Core

**Design Strategies**
- ▶ High-performance single-skin façade with perforated panels
- ▶ Narrow plan depth

**Double-Skin Façade Cavity:**
- ▶ None

**Approximate Percentage of Year Natural Ventilation can be Utilized:**
- ▶ Unpublished

**Percentage of Annual Energy Savings for Heating and Cooling:**
- ▶ 69% compared to a fully air-conditioned German office building (estimated)

**Typical Annual Energy Consumption (Heating/Cooling):**
- ▶ 100 kWh/m² (estimated)

# Highlight Towers Munich, Germany

## Climate

Located in the south of Germany in close proximity to the Alps, the climate in Munich is classified as continental, and experiences annual variation in temperature due to the lack of significant bodies of water nearby. The winter season can be cold and overcast with periods of snow and frost. January is the coldest month, with night-time temperatures falling to −5 °C or lower. Summers are usually pleasantly warm with an average temperature around 23 °C. Occasionally the temperatures can top 28 °C or even 30 °C; however, rainy weather and even thunderstorms are also likely during extreme heat (see Figure 2.9.1).

## Background

The Highlight Towers are part of an urban regeneration project aimed at giving a new life to Parkstadt Schwabing (formerly an industrial wasteland dominated by warehouses) located outside the city center of Munich (see Figure 2.9.2). The towers are part of a larger development which includes two low-rise buildings: an L-shaped, seven-story hotel to the north and a six-story office block to the south. To minimize noise and pollution from the ring road, the two towers are placed in the middle of the site with their narrower façades facing the highway. The two slender skyscrapers (20 meters apart and offset by 15 meters in plan) have parallelogram floor plans with a small footprint that maximizes the amount of daylight penetrating the offices and liberates two-thirds of the site for an open landscaped area (see plan and section, Figures 2.9.3 & 2.9.4).

Glass and steel bridges link both structures at the elevator towers on levels 9, 10, and 20, allowing floors in both towers to be combined, increasing the leasable space on these floors. The bridges are conceived as "clip-on elements" which can be disconnected from the tower and attached to different floors when needed. It is also possible to add extra bridges, catering to various tenant requirements. Vertical circulation elements are placed opposite each other on the

## Climatic Data:[1]

**Location**
▸ Munich, Germany

**Geographic Position**
▸ Latitude 48° 8' N, Longitude 11° 35' E

**Climate Classification**
▸ Temperate

**Prevailing Wind Direction**
▸ West

**Average Wind Speed**
▸ 2.3 meters per second

**Mean Annual Temperature**
▸ 9 °C

**Average Daytime Temperature during the Hottest Months (June, July, August)**
▸ 22 °C

**Average Daytime Temperature during the Coldest Months (December, January, February)**
▸ 4 °C

**Day/Night Temperature Difference During the Hottest Months**
▸ 10 °C

**Mean Annual Precipitation**
▸ 967 millimeters

**Average Relative Humidity**
▸ 72% (hottest months); 84% (coldest months)

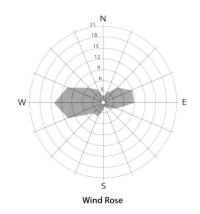

**Wind Rose**

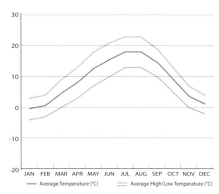

**Average Annual Temperature Profile (°C)**

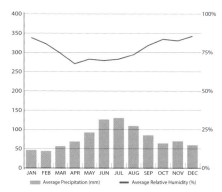

**Average Relative Humidity (%) and Average Annual Rainfall**

▲ Figure 2.9.1: Climate profiles for Munich, Germany.[1]
◀ Figure 2.9.2: Overall view from the southwest. © Murphy/Jahn Architects

[1] The climatic data listed for Munich was derived from the World Meteorological Organization (WMO) and Deutscher Wetterdienst (German Weather Service).

### Plan

*Fresh air is drawn into each floor through pivoting window panels in the exterior façade. Exhaust air passes through a sound-insulated overflow unit into the central corridors, and is then directed through shafts to be discharged from the building.*

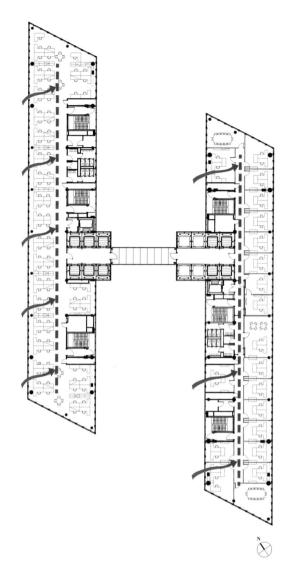

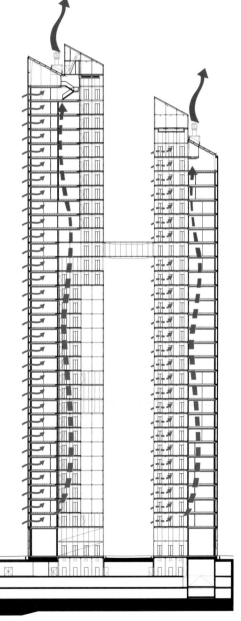

### Section

*Each office is directly ventilated from pivoting panels in the exterior façade. Exhaust air is directed through an overflow unit at the ceiling level and sent to centrally located shafts to be discharged from the building at the roof.*

▲ Figure 2.9.3: Typical office floor plan.
◀ Figure 2.9.4: Building section. (Base drawings © Murphy/Jahn Architects)

▲ Figure 2.9.5: View of the 400 millimeter hinged opening window panels that are spaced around the entire building. © Murphy/Jahn Architects

**The Highlight Towers are naturally ventilated through a single-skin façade with hinged windows protected by perforated, stainless steel panels spaced out across the entire façade.**

inward-facing side of the towers. Removing the elevator cores from the center of the office floor plan and placing them along the periphery of the building creates office space efficiency/ flexibility and exterior views. A variety of office layout options are available: cellular offices with a central corridor; open-plan offices; or a combination of both. Additionally, each floor can also be divided to accommodate two tenants. The ground floor of each tower features a spacious lobby with adjoining showrooms and service facilities.

### Natural Ventilation Strategy

The Highlight Towers are naturally ventilated through a single-skin façade which wraps the entire tower. The typical façade module is 1.35 meters wide and consists of a 950 mm fixed glass panel and a 400 mm hinged window panel (see Figure 2.9.5). The fixed glass panel is triple-glazed and features high-performance and heat-reflecting properties. The narrow operable window features an exterior perforated, stainless steel panel (see Figure 2.9.6) that provides protection from the sun, wind, and rain. This panel also contains soundproofing elements to minimize the noise from the surrounding highway and ring road. The interior side of the hinged window consists of a translucent printed screen pattern applied to heat-insulating double glazing.

The side-hinged windows can be electronically opened inward on each floor to provide individually controlled natural ventilation. When the windows are open, fresh air is drawn into the offices through the fixed perforated panels. Exhaust air passes through a sound-insulated overflow unit into the central corridors, and is then directed through auxiliary rooms and fire protection flaps to the central exhaust shafts to be finally discharged from the building.

A control panel in each office can be used to individually control lighting, ventilation, temperature, and the operability of the windows and blinds. At night, the building management system (BMS) can electronically open the hinged windows to provide night cooling and modulate temperature fluctuations within the building.

▲ Figure 2.9.6: Exterior view of the stainless steel panels that protect the narrow operable windows.
© Murphy/Jahn Architects

windows of the façade can be opened or closed individually to allow occupants direct control over their environment. The operation of these windows can also be controlled automatically by the central BMS to offer night-time ventilation during the hot season.

## Other Sustainable Design Elements

The high-performance, heat-reflecting, triple-glazed windows of the towers provide good insulation and only allow a small amount of solar heat through. Additional solar control is provided by the electronically controlled, highly reflective venetian blinds which are located inside the building (see Figure 2.9.7).

## Analysis – Strengths

▶ The towers have a narrow plan with a depth of 13.5 meters each. This enhances the effectiveness of natural ventilation and daylight performance in the building.

▶ The vertical circulation elements and service facilities (such as staircases, elevator cores and toilets) are placed along the periphery of the building, thus giving more flexibility of internal air movement than with a central core.

▶ The use of a single-skin façade, as opposed to a double-skin façade, minimizes construction cost, saves materials and leasable floor area while optimizing energy consumption and addressing shading, daylighting, visual access, and natural ventilation.

▶ The Highlight Towers has a decentralized displacement ventilation system with fan coil units placed under their raised floor plenums.

## Mixed-Mode Strategy

The Highlight Towers operate without the conventional central mechanical ventilation and air extraction systems. This increases usable floor space without the typical inclusion of vertical ducts. This also allowed for the elimination of a separate mechanical floor. The semi-decentralized system consists of a small plant room (approximately 5m²) located on each floor in each tower.

Mechanical ventilation in the building is provided by a perimeter displacement system. Fresh air is drawn into the building at every floor by two stainless steel elements in the façade. The air is then drawn into a fan coil unit incorporated in the raised floor system. After the air is heated or cooled as required, it travels through the ducts of the raised floor and is distributed to the office spaces through low-velocity displacement outlets.

Displacement ventilation exploits buoyancy rather than fan power as a driving force to draw fresh, cool air into the raised floor and exhaust warm air at the ceiling level, thus reducing energy consumption. In addition, the buildings feature radiant concrete slabs (280 mm thick) with integrated water pipes to provide additional heating and cooling, optimizing the efficiency of the mechanical system. Due to the low humidity levels of Munich, condensation forming on the radiant slabs is not a concern. The building exploits the thermal capacity of concrete and water to regulate internal temperatures and enhance energy performance.

## Interface with the Central Building Management System

The control strategy aims at enhancing individual thermal comfort of tenants in the Highlight Towers. As previously mentioned, the narrow hinged

This eliminates the need for a central plant, a rooftop mechanical penthouse, air extraction systems, shafts, and extensive ductwork. In turn, this saves usable floor area and allows the penthouse to be glazed and rented as a premium space.

▶ Occupants have a high degree of direct control over their environment (through the operation of windows and room control panels). This enhances the individual thermal comfort of tenants in the building.

## Analysis – Considerations

The following areas could be causes for concern if adopting similar strategies in other buildings and should therefore be considered:

▶ Without the use of a double-skin façade to help regulate pressure distribution across the envelope, pressure coefficient values across openings on different floors and sides of the building might vary substantially. This may cause unevenness in airflow patterns on different floors, with some floors better ventilated than others.

▶ When using a cellular office configuration, rooms on one side of the building may experience better ventilation than the other side, depending on the prevailing wind direction.

▶ The perforated panels help protect the towers from the wind, but may not provide adequate protection from excessive drafts or wind-driven rain that typically occur at high levels.

▶ The use of glass with a low visual light transmission will minimize heat gain. However, it may also reduce daylight quality in the building and increase dependence on artificial lighting (which would require additional cooling to address internal head loads created by the artificial lighting).

## Project Team

**Owner/Developer:** Bürozentrum Parkstadt München–Schwabing KG
**Design Architect:** Murphy/Jahn Architects
**Structural Engineer:** Werner Sobek Group
**MEP Engineer:** ENCO Energie-Consulting; Tanssolar
**Environmental Consultant:** Transsolar
**Project Manager:** Schröter + Häubling GbR; Drees & Sommer GmbH
**Main Contractor:** Strabag AG
**Other Consultants:** Ruscheweyh Consultants (Wind Engineering)

**References & Further Reading**

**Books:**

▶ Krauel, J. (ed.) (2008) "Murphy/Jahn: Highlight Munich Business Towers," in *Corporate Buildings*. Links Books: Barcelona, pp. 82–91.

▶ Schmidt, C. (2006) *Highlight Towers*. Braun Publishing: Berlin.

**Journal Articles:**

▶ Anna, S. (2007) "Highlight Business Towers, Munich, Germany, 2004," *World Architecture (China)*, vol. 210, pp. 50–57.

▶ Metz, T. (2006) "Building types study: 855. Offices: culture clash: Highlight Munich Business Towers," *Architectural Record*, vol. 194, no. 3, pp. 154–160.

▲ Figure 2.9.7: Interior view of a typical office space, showing the operable blinds. © Murphy/Jahn Architects

## Project Data:

**Year of Completion**
- ▸ 2005

**Height**
- ▸ 60 meters

**Stories**
- ▸ 17

**Gross Area of Tower**
- ▸ 17,000 square meters

**Building Function**
- ▸ Office

**Structural Material**
- ▸ Concrete

**Plan Depth**
- ▸ 9–12 meters (from central void)

**Location of Plant Floors:**
- ▸ None


## Ventilation Overview:

**Ventilation Type**
- ▸ Natural Ventilation (no mechanical)

**Natural Ventilation Strategy**
- ▸ Cross and Stack Ventilation (connected internal spaces)

**Design Strategies**
- ▸ Rain screen/brise-soleil façade
- ▸ Central (open) atrium
- ▸ Funnel-shaped office spaces

**Double-Skin Façade Cavity:**
- ▸ None

**Approximate Percentage of Year Natural Ventilation can be Utilized:**
- ▸ 100%

**Percentage of Annual Energy Savings for Heating and Cooling:**
- ▸ 100% compared to a conventional air-conditioned building of the same size and typology (assumed)

**Typical Annual Energy Consumption (Heating/Cooling):**
- ▸ 0 kWh/m² (assumed)

# Torre Cube Guadalajara, Mexico

## Climate

Guadalajara is located in a relatively humid sub-tropical climate, featuring dry, mild winters and warm, wet summers. Although the temperature is warm year-round, it experiences a strong seasonal variation in precipitation. There is plenty of sun throughout the year and summer temperatures quickly rise to 30 °C and often approach 35 °C during the months of April and May. The wet season approaches shortly after May, bringing extra moisture, resulting in cooler days and warm nights. While the daytime winter weather tends to be mild, averaging around 25 °C, the temperature quickly drops at night, falling to around 5 °C (see Figure 2.10.1).

## Background

The tower consists of three funnel-shaped, timber-clad office wings cantilevered dramatically between, and from, three concrete cores (see Figure 2.10.2). The three office wings vary in size, typically being 105, 125 and 175 square meters in floor area. Apart from being the primary structural elements, the three cores also contain all the service facilities and vertical circulation elements within the building (e.g., stairwells, elevators, and toilets). The post-tensioned cantilevered slabs allow for open-plan, column-free interiors within the office spaces themselves. The office spaces are a maximum of 12 meters in depth (measuring to the central void) and average approximately 12 meters in width.

The building entrance is on the northeast side, where a grand staircase ascends to the third floor showcasing a plaza-like, open lobby which is secured after hours by a pivoting glass wall. The offices and service cores are arranged around the central open void which functions as a light well and is also an important part of the natural ventilation strategy (see plan and section, Figures 2.10.3 & 2.10.4). In addition to being open at the top, the central void is connected to the exterior by the removal of some office floors which creates sky gardens in each office wing. These sky gardens, or "porches" as they are

## Climatic Data:[1]

**Location**
▶ Guadalajara, Mexico

**Geographic Position**
▶ Latitude 20° 41′N, Longitude 103° 20′W

**Climate Classification**
▶ Temperate (mild seasonal variation)

**Prevailing Wind Direction**
▶ West

**Average Wind Speed**
▶ 4.8 meters per second

**Mean Annual Temperature**
▶ 20 °C

**Average Daytime Temperature during the Hottest Months (April, May, June)**
▶ 32 °C

**Average Daytime Temperature during the Coldest Months (December, January, February)**
▶ 25 °C

**Day/Night Temperature Difference During the Hottest Months**
▶ 19 °C

**Mean Annual Precipitation**
▶ 972 millimeters

**Average Relative Humidity**
▶ 52% (hottest months); 60% (coldest months)

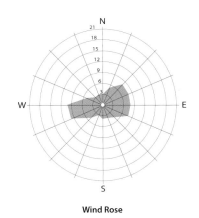

**Wind Rose**

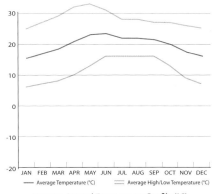

**Average Annual Temperature Profile (°C)**
— Average Temperature (°C)  — Average High/Low Temperature (°C)

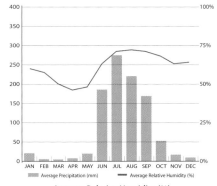

**Average Relative Humidity (%) and Average Annual Rainfall**
▨ Average Precipitation (mm)  — Average Relative Humidity (%)

▲ Figure 2.10.1: Climate profiles for Guadalajara, Mexico.[1]
◀ Figure 2.10.2: Overall view from the east. © Estudio Carme Pinós / Lourdes Grobet

[1] The climatic data listed for Guadalajara was derived from the World Meteorological Organization (WMO) and the National Water Commission of Mexico.

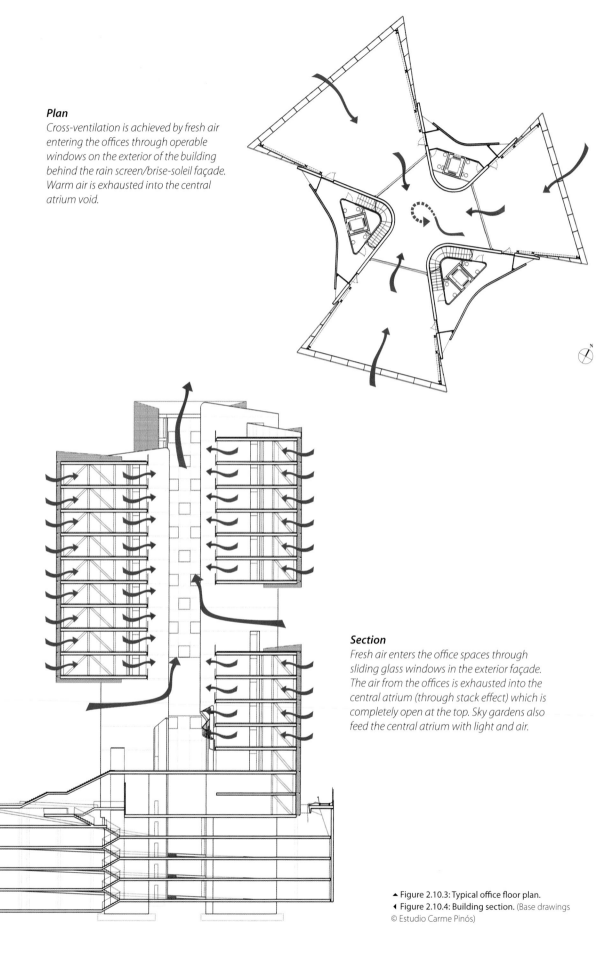

**Plan**

*Cross-ventilation is achieved by fresh air entering the offices through operable windows on the exterior of the building behind the rain screen/brise-soleil façade. Warm air is exhausted into the central atrium void.*

**Section**

*Fresh air enters the office spaces through sliding glass windows in the exterior façade. The air from the offices is exhausted into the central atrium (through stack effect) which is completely open at the top. Sky gardens also feed the central atrium with light and air.*

▲ Figure 2.10.3: Typical office floor plan.
◀ Figure 2.10.4: Building section. (Base drawings © Estudio Carme Pinós)

▲ Figure 2.10.5: View of an interior office space, showing sliding windows with the wooden brise-soleil beyond.
© Estudio Carme Pinós / Lourdes Grobet

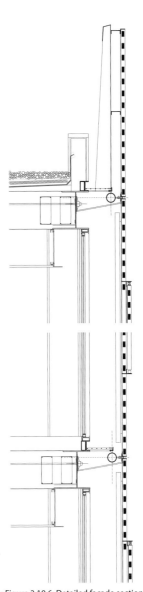

▲ Figure 2.10.6: Detailed façade section through the wooden brise-soleil. © Estudio Carme Pinós

known, serve to bring natural light and ventilation into the central void. The porches also provide generous shade and function as communal terraces for social gatherings.

## Natural Ventilation Strategy

The mild Guadalajara climate allows for natural ventilation throughout the entire year, without reliance on mechanical ventilation, heating, or cooling. The external façades of the three office wings employ an open rain screen/brise-soleil façade and floor-to-ceiling sliding glass windows (see Figures 2.10.5 & 2.10.6). The outer diaphanous screen consists of a wooden latticework made from thin, treated pine battens on a steel frame acting as a brise-soleil, protecting the offices from glare and solar heat gain and acting as protection from falling out of the full-height sliding windows. In addition, this outer screen acts as a partial buffer against wind-driven ventilation into the offices, reducing the speed of the airflow.

The wooden latticework panels can slide horizontally (operated directly by the office occupants), giving a degree

of flexibility to the amount of shade and controlling the flow of air into the offices. As both elements can be manually controlled (the wooden screen and the sliding glazed windows), the occupants have direct control over the amount of sun, light and air entering the office. The intermediate zone between the two façade layers has grated floor panels at each floor, which permit access into the space, but do not prevent vertical airflow within the space.

Air is drawn into the office space through the sliding windows from the façade and exhausted into the atrium through the sliding windows in the inner-facing façade (see Figure 2.10.7). Stack effect in the atrium provides additional uplift, through negative pressure, that pulls air out of the offices to be exhausted at the top of the building. The ventilation strategy of Torre Cube can thus be summarized as a combination of cross-ventilation assisted by significant stack buoyancy in the central atrium.

## Mixed-Mode Strategy

The Torre Cube is entirely dependent on natural ventilation (and solar shading

**The mild Guadalajara climate allowed the architects to design a building that employs natural ventilation throughout the whole year, without any reliance on mechanical ventilation.**

▲ Figure 2.10.7: View from the atrium looking down into the void showing the sliding windows on the inner-facing façade. © Estudio Carme Pinós / Lourdes Grobet

devices) to cool down the building. Thus there is no mechanical ventilation or mixed-mode systems operating. Even during winter, the climate is mild enough that there is no need for heating to warm up the interior. The design does incorporate ductwork to all office spaces that would allow individual tenants to install air-conditioning units for their spaces. However, to date, no tenant has taken this action.

### Interface with the Central Building Management System

The building employs low-tech solutions to its ventilation needs and thus there is no building management system (BMS) system in place. Office occupants manually operate all the windows and wooden lattice screens of the façade. This allows the users direct control over the amount of air and sunlight entering the office spaces.

### Analysis – Strengths

▶ This is a rare example of a high-rise office building (albeit not particularly tall) that is naturally ventilated throughout the entire year with no dependence on air-conditioning. This reduces energy consumption and eliminates the space needed to house mechanical HVAC equipment.

▶ Positioning the services (lifts, staircases, and toilets) in offset structural cores (see Figure 2.10.8) allows for open-plan office spaces with few internal obstructions to affect natural ventilation airflow.

▶ The perimeter of the building is extended, stepping inward in plan where the cores are located, and thus increasing the connection of the office spaces with the external environment. Effectively, each office wing faces the exterior on all four sides, which is beneficial for effective natural ventilation.

▶ The outer diaphanous screen offers both solar protection (reducing glare and excessive solar heat gain) and partially buffers wind-driven ventilation, thus reducing wind speed and making natural ventilation a more viable option.

▶ The stack effect in the central atrium assists cross-ventilation through the office spaces and reduces the dependence on wind-induced ventilation as the atrium will help draw air from the perimeter of all three office wings irrespective of prevailing wind direction.

▶ The central atrium provides daylight at the rear of the office spaces, which allows for the 12-meter-maximum depth.

▶ The three-story sky garden/porches allow for greater air circulation and also function as important social/gathering spaces for office occupants.

▶ By virtue of their form, the fan-shaped office spaces help funnel air into the central void and improve cross-ventilation efficiency between the two window openings.

### Analysis – Considerations

The following areas could be causes for concern if adopting similar strategies in other buildings and should therefore be considered:

▶ If strong winds are blowing from a specific direction, wind becomes the predominant driving force for natural ventilation in that particular office wing. In this case, the office wing facing the prevailing wind may be better ventilated than the others, unless the stack buoyancy in the central atrium is sufficient to draw air from the perimeter windows of those office wings.

▶ The three-story sky gardens within the office wings may reduce the effectiveness of the central atrium as an exhaust device by interrupting the stack buoyancy in the void.

- While enabling the central void, the three off-center structural cores results in duplication of some services (though stairs are only contained in two of the cores), thus affecting the net-to-gross usable floor area efficiency and commercial viability of the project.

- The direct control of the windows may lead to preferential ventilation for some office occupants over others. For example, office workers closer to the external façade may choose to close the windows, reducing ventilation rates to occupants further back in the space.

- With a concrete structure, leaving the ceilings exposed could be used to exploit the material's thermal capacity or to cool down the building at night. This was not done at Torre Cube, which has a suspended gypsum-board ceiling in all office spaces.

- The timber screens would be unlikely to stand up to the greater wind speeds experienced at significant heights, or give a satisfactory longevity for durability, cleaning and maintenance in other climates.

- Existing and future developments could significantly affect the wind patterns around the building. A comprehensive analysis of surrounding existing and future developments would need to be conducted to assess the impact of the urban environment and surrounding structures on the ventilation strategy. CFD and wind tunnel testing will predict airflow patterns around the building to devise control strategies associated with various design alternatives for ventilation, including worst-case scenarios (e.g., a future development that would block prevailing wind directions to the site).

- It should be noted that this particular ventilation strategy works primarily because of the consistent, mild local climate and the design elements were made possible largely because the building is only 16 stories in height. It is unlikely that these strategies would work in climates with larger annual temperature and humidity variations, or with buildings of greater height.

## Project Team

**Owner/Developer:** Cube International
**Design Architect:** Estudio Carme Pinós
**Structural Engineer:** Luis Bozzo Estructuras y Proyectos S.L.
**MEP Engineer:** Anteus Constructora
**Main Contractor:** Anteus Constructora

### References & Further Reading

**Journal Articles:**

- Arriola Clemenz, S. & Pérez-Torres, A. (2006) "Torre Cube, Puerta de Hierro, Guadalajara, Mexico," *On Diseño,* vol. 47, no. 9, pp. 86–107.

- Galiano, L. F. (2005) "Torre Cube, Guadalajara (Mexico)," *AV Monografias,* vol. 115, pp. 18–25.

- Schittich, C. (ed.) (2007) "Torre Cube in Guadalajara, Mexico," *Detail,* vol. 47, no. 9, pp. 962–964, 1076.

▲ Figure 2.10.8: Interior view of one of the three service cores. © Estudio Carme Pinós/Lourdes Grobet

## Project Data:

**Year of Completion**
- ▸ 2007

**Height**
- ▸ 71 meters

**Stories**
- ▸ 18

**Gross Area of Tower**
- ▸ 56,206 square meters

**Building Function**
- ▸ Office

**Structural Material**
- ▸ Concrete

**Plan Depth**
- ▸ 19 meters (between façades)

**Location of Plant Floors:**
- ▸ B1, Roof


## Ventilation Overview:

**Ventilation Type**
- ▸ Mixed-Mode: Zoned / Complementary-Concurrent

**Natural Ventilation Strategy**
- ▸ Wind-Driven Cross-Ventilation

**Design Strategies**
- ▸ Thermal mass
- ▸ Night cooling
- ▸ Voids between cellular offices and ceilings for cross airflow

**Double-Skin Façade Cavity:**
- ▸ None

**Approximate Percentage of Year Natural Ventilation can be Utilized:**
- ▸ 75%

**Percentage of Annual Energy Savings for Heating and Cooling:**
- ▸ 55% compared to a conventional air-conditioned building of the same size and typology (estimated)
- ▸ 15% compared to California Title 24 Energy Code (estimated)

**Typical Annual Energy Consumption (Heating/Cooling):**
- ▸ Unpublished

# San Francisco Federal Building San Francisco, USA

## Climate

Surrounded by water on three sides, the weather in San Francisco is quite moderate. If there is fog, the temperatures tend to drop. During the summer months of July through September, the city is sunny, which burns through any fog by mid-morning and averages a daytime high of 23 °C. Summer is also quite dry and there is rarely any rain from May through September. The city receives its highest rainfall from November through March and winter temperatures remain mild, averaging 14 °C in the daytime (see Figure 2.11.1).

## Background

The San Francisco Federal Building is a slender 18-story tower located in the South-of-Market district (see Figure 2.11.2). The building is fronted by a large open plaza with a free-standing cafeteria pavilion which is a valuable asset to the district. Additionally, the building houses a conference center, fitness center, and daycare center at the lower levels that can be enjoyed by both local residents and employees of the Federal Building.

The narrow rectangular tower, elongated along the northeast–southwest axis, is designed and oriented to take advantage of the prevailing winds from the west (see plan and section, Figures 2.11.3 & 2.11.4). The interior office layout facilitates airflow across the 19-meter-wide floor plate.

## Natural Ventilation Strategy

The prospective for employing natural ventilation with minimal supplementary cooling was determined based on San Francisco's moderate climate and strong prevailing winds. Due to the regulations required of a federal office building, security concerns imposed strict limitations on accessibility and on the openness of the envelope at the base of the building. Therefore, the first five floors are sealed and fully air conditioned and only floors 6–18 benefit from natural ventilation throughout the year.

## Climatic Data:[1]

**Location**
▸ San Francisco, USA

**Geographic Position**
▸ Latitude 37° 47′N, Longitude 122° 26′W

**Climate Classification**
▸ Temperate (mild seasonal variation)

**Prevailing Wind Direction**
▸ West

**Average Wind Speed**
▸ 4.3 meters per second

**Mean Annual Temperature**
▸ 14 °C

**Average Daytime Temperature during the Hottest Months (July, August, September)**
▸ 23 °C

**Average Daytime Temperature during the Coldest Months (December, January, February)**
▸ 14 °C

**Day/Night Temperature Difference During the Hottest Months**
▸ 10 °C

**Mean Annual Precipitation**
▸ 500 millimeters

**Average Relative Humidity**
▸ 69% (hottest months); 69% (coldest months)

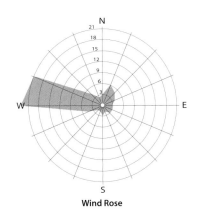

Wind Rose

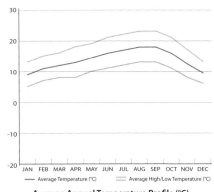

Average Annual Temperature Profile (°C)

— Average Temperature (°C)  ⋯ Average High/Low Temperature (°C)

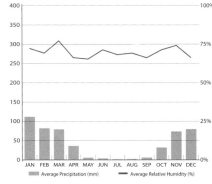

Average Relative Humidity (%) and Average Annual Rainfall

▬ Average Precipitation (mm)  ▬ Average Relative Humidity (%)

▴ Figure 2.11.1: Climate profiles for San Francisco, USA.[1]
◂ Figure 2.11.2: Overall view from the north. © Nic Lehoux

[1] The climatic data listed for San Francisco was derived from the World Meteorological Organization (WMO) and the US National Weather Service.

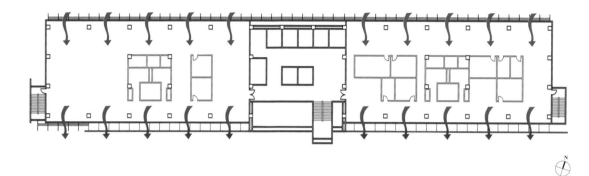

### Plan

Natural ventilation occurs in particular zones of the building. Open-plan office spaces along the building perimeter of levels 6–18 utilize cross-ventilation on a floor-by-floor basis (levels 1–5 are fully sealed and air-conditioned due to security concerns). Cellular offices and conference rooms are mechanically ventilated and located along the central spine of the plan. A 0.5-meter gap to the ceiling is left over the enclosed spaces to allow fresh air to pass over and ventilate the leeward side of the building.

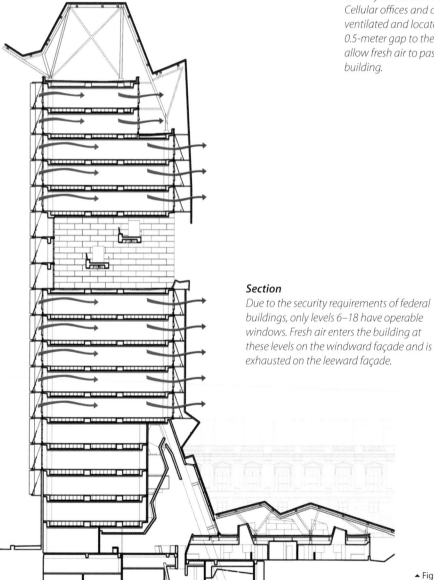

### Section

Due to the security requirements of federal buildings, only levels 6–18 have operable windows. Fresh air enters the building at these levels on the windward façade and is exhausted on the leeward façade.

▲ Figure 2.11.3: Typical office floor plan.
◄ Figure 2.11.4: Building section. (Base drawings © Morphosis)

▲ Figure 2.11.5: Visible at right, manually operable top-hung windows are located at user height, while automated pivoting windows are located above. At left, enclosed spaces leave a 0.5-meter gap to the ceiling to allow air to pass across the floor plate unobstructed. © Nic Lehoux

▲ Figure 2.11.6: Wave-form exposed concrete soffits to increase surface area of the thermal mass. © Nic Lehoux

**A 0.5-meter gap to the ceiling over the central enclosed office spaces and meeting rooms allow cross-ventilated air to pass over and exhaust on the leeward side of the building.**

The natural ventilation strategy is characterized as cross-ventilation on a floor-by-floor basis, enhanced by the design of the building's high-performance façades. Wind enters the building through the operable windows of the windward façade (most typically this is the northwest façade) and is exhausted through windows on the leeward façade. The glazing of both façades features multi-part outward-pivoting, top-hung windows (see Figure 2.11.5). Located at the top of the façade module is a motorized operable window controlled by the BMS. The middle of the façade module contains a manually operable window approximately 50 percent of the façade height. The use of this secondary window opening exclusively controlled by the occupants allows for greater individual control over the ventilation and temperature of each office space.

Conference rooms, enclosed office spaces, bathrooms, and circulation cores are placed along the central spine of the building, allowing for open-plan areas to be located along the perimeter, in close proximity to the windows. The walls of the central spaces do not reach the ceiling, leaving a 0.5-meter gap that permits air to flow from one side of the office space to the other with minimal obstructions (see Figure 2.11.5).

As the temperatures in San Francisco drop at night, the concrete building provides an excellent source for night cooling. The exposed thermal mass is a fundamental component in achieving comfort through natural ventilation only and its effectiveness is enhanced by using upstand structural beams, rather than the standard downstand beams which would interrupt the airflow (McConahey et al. 2002). Additionally, the building features wave-form concrete soffits aimed at increasing the surface area of the thermal mass (see Figure 2.11.6).

## Mixed-Mode Strategy

The ventilation strategy for the San Francisco Federal Building is Mixed-Mode: Zoned. For security reasons the first five floors are sealed and fully air-conditioned. The floors above (levels 6–18) are segmented into two zones. The central spine of each floor, containing fully enclosed rooms and restrooms, is mechanically ventilated and fully air-conditioned. The remaining floor area, consisting of open-plan perimeter office spaces, is ventilated naturally with supplemental mechanical ventilation. However, mechanical air supply registers are only located within the open-plan zones near the mechanically ventilated core and central spine.

The open-plan office spaces above the fifth floor are designed to be naturally ventilated 100 percent of the time (though the actual zone of natural ventilation accounts for approximately 21 percent of the building's total usable area). When heating is required in the naturally ventilated zones, it is provided through a perimeter baseboard system with linear convectors incorporated into the window-wall mullions. Depending on the external temperature, trickle vents located under selected baseboards allow outside air to provide ventilation when occupants would likely close windows due to low exterior temperatures (see Figure 2.11.7).

## Interface with the Central Building Management System

The computerized BMS controls the operation of the building's heating system, lighting levels, and windows in response to the internal and external environmental conditions. The control strategy for the operation of the automatic windows is based on internal air temperature and wind direction by using pressure data to determine the windward side and leeward side. When the wind velocity is very low on warm days and/or when the wind direction is perpendicular to the building's cross-ventilation axis, the BMS will be informed to open the trickle vents (located under selected baseboards). In case of a storm, the system closes the automated windows and opens the trickle vents to provide the minimum amount of required fresh air. At night during warm periods, the BMS opens the motorized windows to flush out the heat built up during the day and bring in night air to cool the exposed concrete structure. More specifically, each floor plate is segmented and monitored by a separate BMS. For each BMS, there are three situations (storm, heating/rain, and mild/cooling) with ten various modes of openness that change based on temperatures, velocities, and pressure differentials.

## Other Sustainable Design Elements

The building has high-performance, double-glazed low-E glass, with an improved solar heat gain coefficient, on both the northwest and southeast façades. Both façades feature shading devices to minimize heat gain in the office spaces.

The glazing of the northwest façade is protected by a series of fixed, translucent vertical sunshades (see Figure 2.11.8), while the glazing of the southeast façade is protected from solar glare and heat gain by a perforated metal sunscreen (see Figure 2.11.9). The screen contains panels which open up in order to give occupants an unobstructed view when the sun has gone around to the other side of the building in the afternoon.

With an average overall ceiling height of roughly four meters, daylight penetrates deep into the office spaces, providing natural illumination for approximately 85 percent of work spaces.

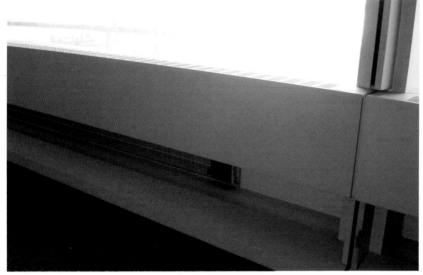

▲ Figure 2.11.7: Trickle vents are located under selected baseboards to allow ventilation when occupants would likely close windows due to low exterior temperatures. © Arup

▲ Figure 2.11.8: Fixed translucent vertical sunshades protect the northwest façade from solar glare and heat gain. © Nic Lehoux

▲ Figure 2.11.9: A perforated metal sun screen protects the southeast façade from solar glare and heat gain. © Nic Lehoux

## Analysis – Performance

The energy targets for the San Francisco Federal Building were defined by the GSA (US General Services Administration). The GSA mandates new buildings to use less than 55,000 Btu of energy per square foot per year (approximately 173 kWh/m²). At the design phase, the building was expected to surpass the GSA's target and Title 24, the stringent California energy code. According to the architect, the building was projected to have an average energy consumption of roughly 37,000 Btu per square foot per year, while levels 6 through 18 were projected to consume less than 28,000 Btu per square foot per year (approximately 116 kWh/m² and 87 kWh/m², respectively) (see Figure 2.11.10).

During the conceptual design phase, the building energy simulation program, EnergyPlus (including the multi-zone, airflow model COMIS) was used to evaluate the performance of several natural ventilation strategies and inform the design of the façade openings and window configurations for optimal thermal comfort. A series of wind tunnel tests were also conducted, including a simulated worst-case scenario of build out on surrounding properties to ensure sufficient natural ventilation could not be compromised. The results concluded that wind-driven ventilation was the most effective strategy and that the building would not require a double-skin façade, resulting in savings of US$1.5 million (Bibb 2008). The results also indicated comfortable conditions could be provided for the

**The open-plan office spaces above the fifth floor are designed to be naturally ventilated 100 percent of the time (though this area of natural ventilation accounts for approximately 21 percent of the building's total usable area).**

# The CFD results informed the shape of the window mullions by adding an inflow deflector which serves to deflect incoming air away from the desk and seating areas of the windward side of the building.

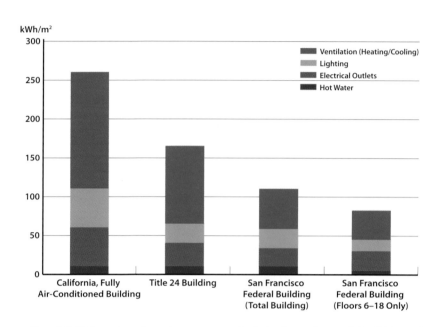

▲ Figure 2.11.10: Comparison of annual energy consumption for heating, cooling, ventilation, lighting, and hot water for the San Francisco Federal Building. (Source: Morphosis Architects)

majority of occupiable hours in the year, if night cooling is provided in periods of hot weather (Carrilho da Graça et al. 2004).

For fine-tuning the selected strategy, detailed CFD simulations predicted cross-flow ventilation airflow patterns, indoor air velocities, indoor temperature distribution and ventilation flow rates associated with different combinations of window openings and exterior conditions such as wind speeds. The CFD results informed the configuration and sizing of windows, their associated control algorithm, and the interior layout of the office furniture. Furthermore, these studies finalized the shape of the window mullions by adding an inflow deflector (see Figure 2.11.11) which serves to deflect incoming air away from the desk and seating areas of the windward side of the building (Bibb 2008).

## Analysis – Strengths

▶ The interior layout of the office space is designed to maximize airflow. CFD simulations confirm that locating open-plan offices along the perimeter and enclosed spaces along the central spine can enhance natural ventilation. Leaving a 0.5-meter gap to the ceiling above the enclosed spaces in the central area allows cross-ventilation between windward and leeward faces.

▶ The narrow floor plate and orientation maximize natural airflow for optimal cooling and ventilation.

▶ The building's thermal mass, optimization of the façade, and the effective solar control strategy reduce the cooling load of the building, maximize the potential for natural ventilation, and eliminate the dependence on air-conditioning.

▶ The use of upstand beams instead of the more commonly used downstand beams allow for a raised floor plenum which offers under-floor air distribution for displacement ventilation, flexibility in routing of telecommunication and electrical power conduit, and ease of access to main piping routes. Furthermore, this contributes to the benefit of night cooling by allowing airflow to cool the exposed soffit of the slab.

▶ The use of an inflow deflector on the glazing mullions enhances occupant comfort and the effectiveness of natural ventilation. The effect of adding a deflector demonstrates how small improvements in window geometry (informed by CFD simulations) can have a significant impact on indoor flow distribution and the effectiveness of natural ventilation, especially when wind is a major driving force (Haves et al. 2004).

## Analysis – Considerations

The following areas could be causes for concern if adopting similar strategies in other buildings and should therefore be considered:

▸ Incorrect user control and/or the lack of sufficient education on the operation of the building can lead to poor building performance. For instance, when the BMS tries to make optimal use of the available thermal capacity of the cooled concrete slab on hot days, windows opened by the users can result in an increase in indoor temperatures and thermal discomfort (Carrilho da Graça et al. 2004).

▸ Without a proper control strategy accounting for variations in user behavior, occupants on one side of the building can negatively affect climate control on the other side. This also applies if a proper methodology for the configuration and sizing of windward versus leeward openings is not implemented.

▸ This strategy could risk a building overheating during a sequence of hot summer days, when stored thermal capacity (night cooling) and increased natural ventilation rate is inappropriate due to high ambient temperatures.

▸ Direct wind-driven air into the office spaces (i.e., not via an intermediate space such as a double-skin façade) may result in high air change rates, movement of paper, and occupant discomfort.

## Project Team

**Owner/Developer:** US General Services Administration
**Design Architect:** Morphosis
**Associate Architect:** Smith Group
**Structural Engineer:** Arup
**MEP Engineer:** Arup
**Project Manager:** Hunt Construction Group
**Main Contractor:** Dick Corporation; Morganti General Contractors
**Other Consultants:** Lawrence Berkeley National Laboratory (Natural Ventilation Modeling); Curtain Wall Design & Consulting, Inc. (Curtain Wall Consultant)

**References & Further Reading**

**Books:**

▸ Carter, B. (2008) *GSA / Morphosis / Arup: Integrated Design – San Francisco Federal Building*. Buffalo Workshop: New York.

▸ McConahey, E., Haves, P. & Christ, T. (2002) "The integration of engineering and architecture: a perspective on natural ventilation for the new San Francisco Federal Building," in *2002 ACEEE Summer Study on Energy Efficiency in Buildings*. American Council for an Energy Efficient Economy: Washington, DC, pp. 239–252.

▸ Wood, A. (ed.) (2008) *Best Tall Buildings 2008: CTBUH International Award Winning Projects*. Elsevier Inc./Architectural Press: Burlington, pp. 30–31.

**Journal Articles:**

▸ Carrilho da Graça, G., Linden, P. F. & Haves, P. (2004) "Design and testing of a control strategy for a large, naturally ventilated office building," *Building Services Engineering Research and Technology Journal*, vol. 25, no. 3, pp. 223–239.

▸ Haves, P., Linden, P. F. & Carrilho da Graça, G. (2004) "Use of simulation in the design of a large, naturally ventilated office building," *Building Services Engineering Research and Technology Journal*, vol. 25, no. 3, pp. 211–221.

**Reports:**

▸ Bibb, D. (ed.) (2008) "The case for sustainability," in *Sustainability Matters*. US General Services Administration: Washington, DC, pp. 8–33.

▸ Brager, G., Borgeson, S. & Lee, Y. (2007) *Summary Report: Control Strategies for Mixed-Mode Buildings*. Center for the Built Environment (CBE), University of California, Berkeley, pp. 34–38.

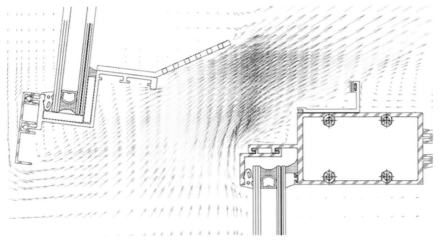

▲ Figure 2.11.11: CFD simulation of inflow behavior with the flow deflector for occupant-controlled windows.
(Source: University of California at San Diego)

## Project Data:

**Year of Completion**
- ▶ 2008

**Height**
- ▶ 115 meters

**Stories**
- ▶ 22

**Gross Area of Tower**
- ▶ 64,567 square meters

**Building Function**
- ▶ Office

**Structural Material**
- ▶ Concrete

**Plan Depth**
- ▶ 11.5 meters (from central core)

**Location of Plant Floors:**
- ▶ B1, Roof

## Ventilation Overview:

**Ventilation Type**
- ▶ Mixed-Mode: Complementary-Concurrent

**Natural Ventilation Strategy**
- ▶ Cross and Stack Ventilation (connected internal spaces)

**Design Strategies**
- ▶ Double-skin façades
- ▶ Segmented atria/sky gardens act as a thermal buffer zone
- ▶ 115-meter solar chimney

**Double-Skin Façade Cavity:**
- ▶ Depth: 1.3 meters
- ▶ Horizontal Continuity: Fully Continuous (along entire length of façade)
- ▶ Vertical Continuity: 4 meters (floor-to-floor)

**Approximate Percentage of Year Natural Ventilation can be Utilized:**
- ▶ 35%

**Percentage of Annual Energy Savings for Heating and Cooling:**
- ▶ 73% compared to a fully air-conditioned Manitoba office building (measured)
- ▶ 81% compared to Canadian MNECB energy code (measured)

**Typical Annual Energy Consumption (Heating/Cooling):**
- ▶ 39 kWh/m² (measured)

# Manitoba Hydro Place Winnipeg, Canada

## Climate

Winnipeg experiences extreme variations in climate, with a temperature swing between a −35 °C extreme in winter and a +35 °C extreme in summer. From December through February the temperature typically remains below freezing, with an average daytime high of −10 °C. Despite the harsh winter, Winnipeg is known for year-round sunny weather. Snowfall is regular, with an annual average of 1,150 millimeters and can continue through springtime. By April, temperatures begin peaking at 10 °C. Summertime experiences hot weather with plenty of sunshine and the temperature often peaks at 30 °C between July and August (see Figure 2.12.1).

## Background

Careful consideration was given to the new location for the Manitoba Hydro headquarters, including a study on proximity to public transport options for the 1,800 employees. The project was seen as an opportunity to build up downtown Winnipeg and the local transit system. Prior to relocating from the suburbs, 95 percent of the employees drove to work. After relocation and a corporate incentive program, 68 percent of the employees use public transport to get to and from work (Linn 2010). In addition to the city's local restaurants and shops benefitting from 1,800 new patrons, the new building offers a three-story, light-filled galleria which is open to the public.

The mass of Manitoba Hydro Place consists of two converging 18-story office wings separated by a service core, resting on a three-story podium (see Figure 2.12.2). The two column-free office blocks face west and east-northeast respectively, with north- and south-facing atria fusing the two masses together. Private office spaces such as workstations and glass-enclosed meeting areas are organized into "neighborhoods" around each atrium. The form and mass of the towers were generated in response to solar orientation, prevailing wind conditions, and other unique climatic conditions that characterize the city of Winnipeg.

## Climatic Data:[1]

**Location**
▸ Winnipeg, Canada
**Geographic Position**
▸ Latitude 49° 54' N, Longitude 97° 7'W
**Climate Classification**
▸ Cold
**Prevailing Wind Direction**
▸ South
**Average Wind Speed**
▸ 4.7 meters per second
**Mean Annual Temperature**
▸ 3 °C
**Average Daytime Temperature during the Hottest Months (June, July, August)**
▸ 25 °C
**Average Daytime Temperature during the Coldest Months (December, January, February)**
▸ −10 °C
**Day/Night Temperature Difference During the Hottest Months**
▸ 13 °C
**Mean Annual Precipitation**
▸ 514 millimeters
**Average Relative Humidity**
▸ 69% (hottest months); 77% (coldest months)

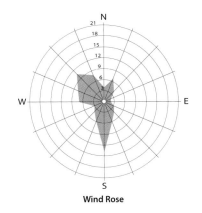

**Wind Rose**

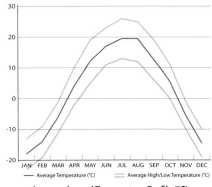

**Average Annual Temperature Profile (°C)**

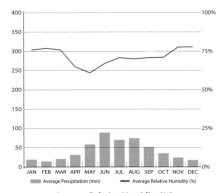

**Average Relative Humidity (%) and Average Annual Rainfall**

▲ Figure 2.12.1: Climate profiles for Winnipeg, Canada.[1]
◀ Figure 2.12.2: Overall view from northwest. © KPMB Architects

[1] The climatic data listed for Winnipeg was derived from the World Meteorological Organization (WMO) and the Meteorological Service of Canada.

## Plan

*Fresh air is drawn in through the south atrium. The air then passes into the raised floor distribution plenum and is distributed along the building perimeter. During the shoulder seasons, supplemental natural ventilation can be provided through the west and northeast double-skin façades. The air is then drawn through glass louvers to the north atrium where stack effect in the solar chimney pulls air from the north atrium and exhausts the building.*

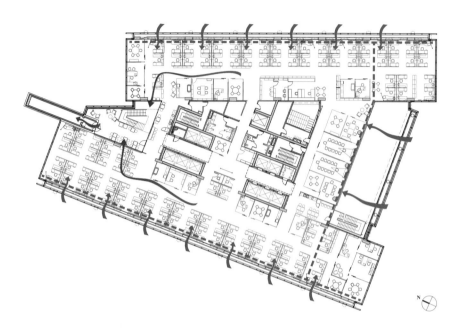

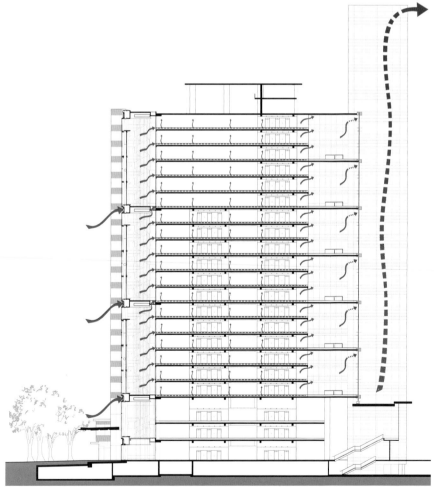

## Section

*During the summer and shoulder seasons, louvered bands on the south atria façade allow fresh air to enter the sky gardens, which is then distributed to offices via under-floor plenums. Stack effect within the northern solar chimney pulls fresh air across the office floors to exhaust it vertically. During the winter season, dampers within the chimney close and fans draw in warm, exhaust air to recapture heat for the southern winter gardens. Concurrently, cool air is drawn into the south atria through outer mechanical units. Recovered heat and solar gain warm fresh air before it enters the office floors.*

▲ Figure 2.12.3: Typical office floor plan.
◀ Figure 2.12.4: Building section. (Base drawings © KPMB Architects)

▲ Figure 2.12.5: South façade showing the air intake grilles between the three stacked, six-story-high atria.
© KPMB Architects/Gerry Kopelow

The building plan was designed with an "A" configuration, splaying open to form a south atrium to capture Winnipeg's abundant winter sunlight and consistent southerly prevailing winds for passive solar heating and natural ventilation. The two wings converge at the north end and feature a 115-meter-tall solar chimney which runs continuously from ground level to several stories beyond the roof (see plan and section, Figures 2.12.3 & 2.12.4).

### Natural Ventilation Strategy

The building orientation and massing are key strategies to naturally ventilate the building. As the building plan opens up to the south, three stacked atria take full advantage of sunlight and southerly prevailing winds. Cross-ventilation draws fresh air across the office spaces and stale air is exhausted in the north atria solar chimney.

Each of the three stacked, six-story-high sky gardens draw in fresh air through louvered air inlet bands at each sky garden (see Figure 2.12.5) (because the air is then drawn in through the building, it is rarely exhausted from these atria directly, however, smoke exhaust fans and vents are located at the western edge of each band). During the winter, fresh air enters through air-handling units and is pre-heated by the heat-recovery unit and solar heat gain in the south atria. The incoming air is humidified by a water feature that flows from the ceiling to the floor (see Figure 2.12.6). During the summer, the water feature is chilled to the point where it can dehumidify fresh air entering the atria. Additional dehumidification and cooling/conditioning is achieved by fan coils which distribute air to the offices through an under-floor displacement ventilation system. Opaque panels project out from the face of the inner façade enclosing the southern atrium

**The building plan was designed with an "A" configuration, splaying open to form a south atrium to capture Winnipeg's abundant winter sunlight and consistent southerly prevailing winds for passive solar heating and natural ventilation.**

▲ Figure 2.12.6: Drawing of the south atrium showing the water feature.
© KPMB Architects/Bryan Christie Designs

▲ Figure 2.12.7: South atrium showing the opaque spandrel panels on the inner façade that inlet fresh air to the under-floor distribution to the offices.
© KPMB Architects/Eduard Hueber

**The incoming air is humidified by a water feature that flows from the ceiling to the floor. During the summer, the water feature is chilled to the point where it can dehumidify fresh air entering the atria.**

(see Figure 2.12.7) to draw air into the floor levels.

Supplementing ventilation from the south atria, a west and northeast double-skin façade admit additional fresh air into office spaces when external temperatures allow (see Figures 2.12.8 & 2.12.9). The two façades consist of two, low-iron glass curtain walls (double-glazed exterior layer and single-glazed interior layer) with a 1.3-meter cavity that contains motorized/operable blinds. Exterior, motorized glazing panels are auto-mated and connected to the BMS. They are automatically opened when the outdoor air temperature is greater than 5 °C and wind speeds do not exceed 20 m/s. They are automatically closed when temperatures are below 0 °C to avoid condensation on the curtain wall gaskets. Manually operated windows

are located on the inner façade and completely controlled by the occupants. These operable windows are located at desk height to keep the air delivered at the occupant's level. As fresh air travels horizontally through the office spaces, building occupants, equipment, and other internal heat sources cause the air to rise. The air is drawn toward the north atria through louvered glazing at the ceiling level (see Figure 2.12.10) by the underpressure created by the solar chimney (due to stack effect).

The north atria consist of stacked, three-story sky gardens (see Figure 2.12.11). Exhaust air is drawn into this atrium and exits to the solar chimney through louvered dampers at the top of each atrium. The solar chimney relies on natural stack effect to exhaust air from the building during summer and

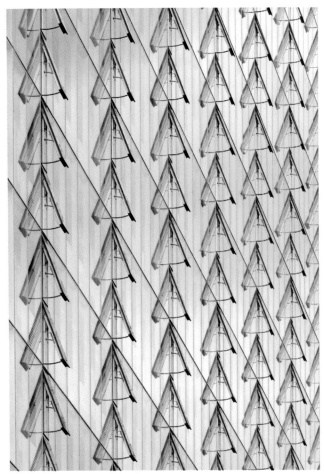

▲ Figure 2.12.8: Operable glazing panels on the exterior layer of the double-skin façade. © KPMB Architects/Tom Arban

▲ Figure 2.12.9: View of the northeast double-skin façade. © KPMB Architects/Eduard Hueber

mid-seasons. In winter, a heat-recovery unit at the bottom of the solar chimney draws in exhaust air to warm the parking garage and preheat the south atria. Dampers in the solar chimney maintain a constant pressure difference from the offices into the chimney at each level to ensure that an even amount of exhaust air is taken from each floor (e.g., the dampers are more closed at the lower levels where stack effect is strongest). To ensure effective stack effect during cool summer nights or cloudy days, sand-filled pipes installed behind the glass at the top of the chimney absorb and store solar heat, inducing the rising of air (see Figure 2.12.12).

In contrast to a conventional North American office building, where approximately 80 percent of supplied air is re-circulated and only 20 percent is fresh, Manitoba Hydro Place supplies 100 percent fresh air every day throughout the year. This is available, regardless of outside temperatures, due to the highly efficient heat-recovery system.

## Mixed-Mode Strategy

The building operates as a Complementary-Concurrent building. The mechanical system runs during the winter and summer to provide supplemental heating and cooling with the majority of the mechanical systems switched off during the shoulder seasons to allow for natural ventilation. During the colder winter months, a heat-recovery system extracts heat from exhaust air in the solar chimney to preheat incoming cold air in the south atria. If supplementary heating or cooling is needed, under-floor displacement ventilation (on a floor-by-floor basis) is tempered from fan coil units located in the raised floor system.

Radiant ceilings provide heating or cooling depending on the season and extend into the double-skin cavity to ensure temperatures do not fall below 3 °C. The heat or cold for the radiant system is generated by a geothermal system and reversible heat pump (chiller). Due to the advantage of the radiant system, high-temperature water during the heating season (maximum 35 °C) and low-temperature water during the cooling season (minimum 18 °C) allow for the efficient operation of a reversible heat pump. The system can operate in a free cooling mode when cold water from the geothermal field is directly used and the reversible heat pump is not needed to run.

▲ Figure 2.12.10: Glass partitions to north atrium showing louvered vents that exhaust office spaces.
© KPMB Architects

▲ Figure 2.12.11: View into a three-story north atrium which exhausts air into the solar chimney.
© KPMB Architects/Tom Arban

## Other Sustainable Design Elements

The building features insulating double-skin façades along the west and northeast side of the tower. Automated louver shades control glare and solar heat gain. The concrete structure is designed to create a thermal mass that moderates the impact of extreme temperature swings. Exposed radiant concrete ceilings act as internal heat exchangers, maintaining a comfortable temperature year-round. A geothermal system supplies water to the radiant ceilings. The building orientation supports passive solar gains to condition the space and the high ceilings increase natural daylighting. The highly efficient heat-recovery system reduces energy consumption by preheating intake air and the displacement ventilation ensures continuous fresh air supply

## Analysis – Performance Data

Manitoba Hydro Place required an extensive commissioning process to ensure the building met energy efficiency and occupant comfort targets. This commissioning phase was followed by an optimization phase, when building systems were studied and modified for maximum comfort and sustainability. An in-depth measurement and verification plan to IPMVP standards (International Performance Measurement & Verification Protocol) was developed by the building energy management engineer. This plan will utilize data collected by the BMS to develop an as-built building model. These tools will be utilized by an Energy Advisory Team to ensure that Manitoba Hydro Place will meet or exceed its energy targets.

As of 2011, the building has significantly reduced heating loads, which were measured at 29 kWh/m$^2$ (cooling loads were only 10 kWh/m$^2$). A typical office

## Interface with the Central Building Management System

The building management system (BMS) of Manitoba Hydro Place is unique due to the sheer number of observation points. With more than 25,000 observation points and two local on-site weather stations, the building performance can be monitored in close detail. The BMS automatically controls the operation of windows and vents, shading devices, temperature, ventilation, and radiant ceilings. For example, when heat builds up in the cavity of the double-skin façade, the BMS automatically drops the shades to reduce further solar heat gain and opens the windows of the outer skin to vent the façade cavity. Conversely, a custom computer interface enables employees to control/override aspects of lighting and solar shading. By leveraging information technology to monitor and control the building, while enabling the individual user control of their immediate environment, a more responsive and productive work environment is created.

▲ Figure 2.12.12: Detail view of solar chimney crown. © KPMB Architects

In contrast to a conventional North American office building that supplies only 20 percent fresh air, Manitoba Hydro Place supplies 100 percent fresh air regardless of outside temperatures.

tower in this climate would require 250–300 KWh/m² for heating. This dramatic reduction is due to passive solar heat gain, thermal mass which allows night-time and weekend shutdowns of the heating system, and an effective building envelope. The double-skin façades are passively maintained to 10 °C during the peak heating season, which significantly reduces heat loss through the envelope. According to the climate engineer, once the ground-source heat loop is commissioned, the energy usage for heating is expected to meet its target set at 15.2 kWh/m².

Energy savings for office lighting was projected to be 65 percent. With the combination of high-quality glazing, a narrow floor plate, and advanced lighting fixtures, preliminary data support this savings target. Current lighting loads are at 56 percent savings, with an uncommissioned system. The climate engineer has confirmed that Manitoba

Hydro Place is meeting its building design energy target and is expected to significantly exceed it at the conclusion of the optimization phase.

A typical office tower in Canada operates with an annual energy consumption range of 400–550 kWh/m². A typical Manitoba office tower utilizes 495 kWh/m². Over the past decade that has been reduced to 325 kWh/m² due to Manitoba Hydro initiatives as a utility provider. New performance guidelines based on the Model National Energy Code for Buildings (MNECB) target an annual energy consumption of 260 kWh/m². Current consumption patterns at Manitoba Hydro Place demonstrate a projected total energy use of 88 kWh/m². This 66 percent energy savings surpasses the targeted 60 percent energy savings of the MNECB code (Sampson 2010).

### Analysis – Strengths

▸ Manitoba Hydro Place presents a successful model for a naturally ventilated high-rise building in an extremely cold, dry winter climate (with a relatively high amount of sun-hours) . The use of south-facing atria or winter gardens to preheat incoming fresh air is ideal for natural ventilation in cold climates.

▸ The orientation, configuration and layout of the building are very responsive to Winnipeg's local climatic conditions and well integrated with the tower's passive design strategies. The building is oriented to take advantage of the southern prevailing winds and solar heat to warm incoming cool air.

▸ The use of atria/sky gardens creates a microclimate, tempers incoming

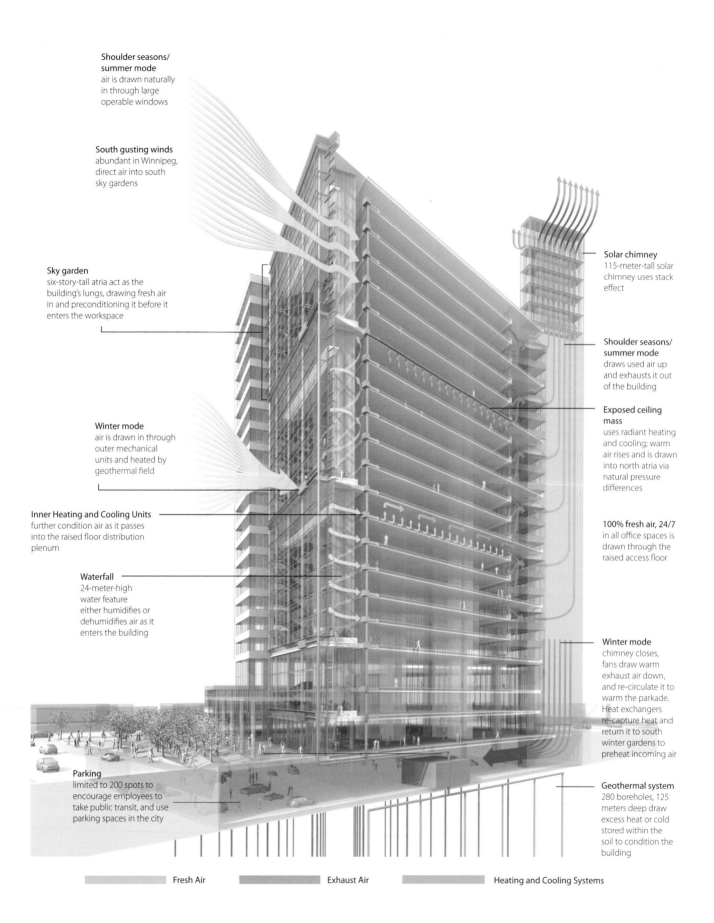

**Shoulder seasons/ summer mode**
air is drawn naturally in through large operable windows

**South gusting winds**
abundant in Winnipeg, direct air into south sky gardens

**Sky garden**
six-story-tall atria act as the building's lungs, drawing fresh air in and preconditioning it before it enters the workspace

**Winter mode**
air is drawn in through outer mechanical units and heated by geothermal field

**Inner Heating and Cooling Units**
further condition air as it passes into the raised floor distribution plenum

**Waterfall**
24-meter-high water feature either humidifies or dehumidifies air as it enters the building

**Parking**
limited to 200 spots to encourage employees to take public transit, and use parking spaces in the city

**Solar chimney**
115-meter-tall solar chimney uses stack effect

**Shoulder seasons/ summer mode**
draws used air up and exhausts it out of the building

**Exposed ceiling mass**
uses radiant heating and cooling; warm air rises and is drawn into north atria via natural pressure differences

**100% fresh air, 24/7**
in all office spaces is drawn through the raised access floor

**Winter mode**
chimney closes, fans draw warm exhaust air down, and re-circulate it to warm the parkade. Heat exchangers re-capture heat and return it to south winter gardens to preheat incoming air

**Geothermal system**
280 boreholes, 125 meters deep draw excess heat or cold stored within the soil to condition the building

Fresh Air          Exhaust Air          Heating and Cooling Systems

▲ Figure 2.12.13: Overview of Manitoba Hydro Place's environmental strategy. © KPMB Architects/Bryan Christie Designs

fresh air for natural ventilation, and mitigates unwanted drafts.

▶ The use of a water feature to humidify or dehumidify the incoming fresh air facilitates natural ventilation in challenging climates with extreme temperatures, humidity and/or dryness.

▶ The solar chimney at the north end of the building induces the flow of air across the office spaces and improves the efficiency of natural ventilation when the prevailing south wind is weak.

▶ Filling the top of the solar chimney with a high thermal capacity material, such as sand, maintains stack effect during cool summer nights and enhances the effectiveness of night-purge ventilation. The solar heated, sand-filled pipes ensure a sufficient temperature and pressure difference is maintained in the chimney to exhaust the building.

▶ Extending the solar chimney several stories beyond the roof level of the high-rise building serves to enhance the stack effect and ensure that the top floors of the towers are sufficiently ventilated.

▶ Locating open-plan office layouts near the building perimeter with private meeting spaces and service areas toward the core complements and optimizes the natural ventilation strategy.

▶ The segmentation of the south-facing atrium reduces the risks of large pressure differentials at the top and bottom of the atrium. This minimizes the risk of top floors receiving significantly warmer air than those at the bottom and avoids the complexity associated with sizing the openings adjoining the office spaces to the atrium.

## Analysis – Considerations

The following areas could be causes for concern if adopting similar strategies in other buildings and should therefore be considered:

▶ Without proper control strategies such as shading and venting, the south-facing atria may be subject to overheating and therefore potentially overheat the fresh air for natural ventilation.

▶ Condensation and ice could occur at fresh air intake openings (windows and vents) if opened during extremely cold weather conditions. In this climate, there is a potential risk of snow blocking the ventilation inlets.

▶ Future developments, such as the erection of a high-rise to the south of the site, might cast a shadow on the sky gardens and have a negative impact on the performance of the natural ventilation strategy.

## Project Team:

**Owner/Developer:** Manitoba Hydro
**Design Architect:** Kuwabara Payne McKenna Blumberg Architects
**Associate Architect:** Smith Carter Architects and Engineers
**Structural Engineer:** Halcrow Yolles; Croslier Kilgour & Partners Ltd.
**MEP Engineer:** AECOM
**Climate Engineer:** Transsolar
**Main Contractor:** PCL Constructions Eastern; J.D. Strachan Construction Ltd.
**Other Consultants:** RWDI (wind engineering); Prairie Architects (LEED consultant)

**References & Further Reading**

**Books:**

▶ Kuwabara, B., Auer, T., Gouldsborough, T., Akerstream, T. & Klym, G. (2009) "Manitoba Hydro Place: integrated design process exemplar," in Demers, C. & Potvin, A. (eds.) *Proceedings of PLEA 2009.* Les Presses de l'Université Laval: Quebec City, pp. 551–556.

▶ Moe, K. (2008) *Integrated Design in Contemporary Architecture.* Princeton Architectural Press: Princeton, pp. 18–23.

▶ Wood, A. (ed.) (2010) *Best Tall Buildings 2009: CTBUH International Award Winning Projects.* Routledge: New York, pp. 20–27.

**Journal Articles:**

▶ Chodikoff, I. (2006) "Award of excellence: Manitoba Hydro head office," *Canadian Architect,* vol. 51, no. 12, pp. 32–35.

▶ Dassler, F. H. (2007) "Extreme Randbedungen: interview with Bruce Kuwabara and Thomas Auer," *XIA Intelligente Architetektur,* vol. 58, no. 1, pp. 18–25.

▶ Linn, C. (2010) "Cold comfort," *Green Source,* March/April, pp. 52–57.

▶ Sampson, P. (2010) "Climate-controlled," *Canadian Architect,* vol. 55, no. 1, pp. 16–22.

▶ Slavic, D. (2008) "IBS Award 2008," *XIA International Magazine,* vol. 8, no. 2, pp. 11–17.

## Project Data:

**Year of Completion**
▸ 2010

**Height**
▸ 56 meters

**Stories**
▸ 14

**Gross Area of Tower**
▸ 22,300 square meters

**Building Function**
▸ Office

**Structural Material**
▸ Composite

**Plan Depth**
▸ 6.3 meters (from central core)

**Location of Plant Floors:**
▸ Roof


## Ventilation Overview:

**Ventilation Type**
▸ Mixed-Mode:
   Complementary-Concurrent

**Natural Ventilation Strategy**
▸ Wind-Driven Cross-Ventilation
▸ Stack Ventilation via Vertical Shaft in
   Core

**Design Strategies**
▸ Double-skin façade
▸ "Pressure Ring" which maintains a ring
   of consistent positive pressure
▸ Aerodynamic external form

**Double-Skin Façade Cavity:**
▸ Depth: 700 mm
▸ Horizontal Continuity: Fully
   Continuous (around entire perimeter)
▸ Vertical Continuity: 3.7 meters
   (floor-to-floor)

**Approximate Percentage of Year Natural
Ventilation can be Utilized:**
▸ 60%

**Percentage of Annual Energy Savings for
Heating and Cooling:**
▸ 84% compared to a fully air-
   conditioned German office building
   (estimated)

**Typical Annual Energy Consumption
(Heating/Cooling):**
▸ 50 kWh/m² (estimated)

# KfW Westarkade Frankfurt, Germany

## Climate

Frankfurt is located in a warm temperate climate which experiences relatively mild weather all year round. Although fairly uncommon, short-lived extremes occasionally occur. Summers are typically warm and sunny with a chance of light rain occurring. On the sunniest of days, the temperature often rises above 24 °C, even topping 30 °C on occasion, and drops to around 15 °C in the evenings. During the winter season, January is characteristically the coldest month. Daytime temperatures hover around 4 °C and at night drop just below freezing, when a light snow is likely (see Figure 2.13.1).

## Background

KfW Westarkade was built as an expansion to the existing KfW Bankengruppe headquarters complex and provides additional office space for 700 employees. KfW Bankengruppe, a government-owned German development bank, finances investments in domestic energy efficiency, energy conservation, and environmental protection initiatives. Since the bank has played a leading role in defining some of Germany's strict energy policies, it is no surprise they mandated a green building, running on less than 100 kWh/m² of primary energy[2] annually. The client also required a scheme that aimed at preserving views and daylight for occupants of the bank's existing cluster of buildings.

In an effort to integrate the tower into its context but also to save energy, the tower was designed as a streamlined flowing form (see Figure 2.13.2) that responds to the direction of prevailing winds and the sun's daily and annual path. The airfoil-shaped tower rests on a curved four-story podium which connects to adjacent KfW buildings on multiple levels. In addition, the podium is integrated with its surrounding context by defining a street edge and creating a courtyard that flows into the adjacent public park, Palmengarten.

## Climatic Data:[1]

**Location**
▸ Frankfurt, Germany

**Geographic Position**
▸ Latitude 50° 7′N, Longitude 8° 41′E

**Climate Classification**
▸ Temperate

**Prevailing Wind Direction**
▸ South-southwest

**Average Wind Speed**
▸ 4 meters per second

**Mean Annual Temperature**
▸ 10 °C

**Average Daytime Temperature during the Hottest Months (June, July, August)**
▸ 24 °C

**Average Daytime Temperature during the Coldest Months (December, January, February)**
▸ 4 °C

**Day/Night Temperature Difference During the Hottest Months**
▸ 11 °C

**Mean Annual Precipitation**
▸ 621 millimeters

**Average Relative Humidity**
▸ 53% (hottest months); 76% (coldest months)

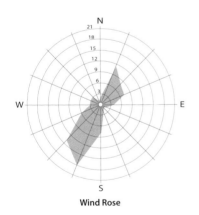

**Wind Rose**

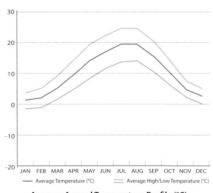

**Average Annual Temperature Profile (°C)**
— Average Temperature (°C)    Average High/Low Temperature (°C)

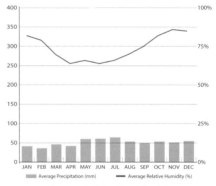

**Average Relative Humidity (%) and Average Annual Rainfall**
Average Precipitation (mm)    — Average Relative Humidity (%)

▴ Figure 2.13.1: Climate profiles for Frankfurt, Germany.[1]
◂ Figure 2.13.2: Overall view from the south. © Jan Bitter

---

[1] The climatic data listed for Frankfurt was derived from the World Meteorological Organization (WMO) and Deutscher Wetterdienst (German Weather Service).

[2] Primary energy is measured at the natural resource level, including losses from the processes of extraction of resources, transformation, and distribution. It expresses the building's full load on the environment.

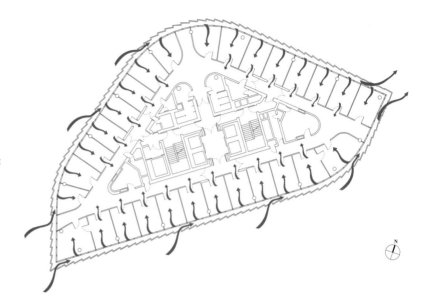

## Plan

The offices are ventilated directly through a double-skin façade system, maintained (by the BMS) at a slightly higher pressure than the interior. Outward-pivoting colored glass panels in the outer façade and side-hung windows in the inner façade allow fresh air into the offices. Cross-ventilation then draws air from the offices through the corridor and into a lower pressurized core. Stack effect in this core exhausts air at the top of the building.

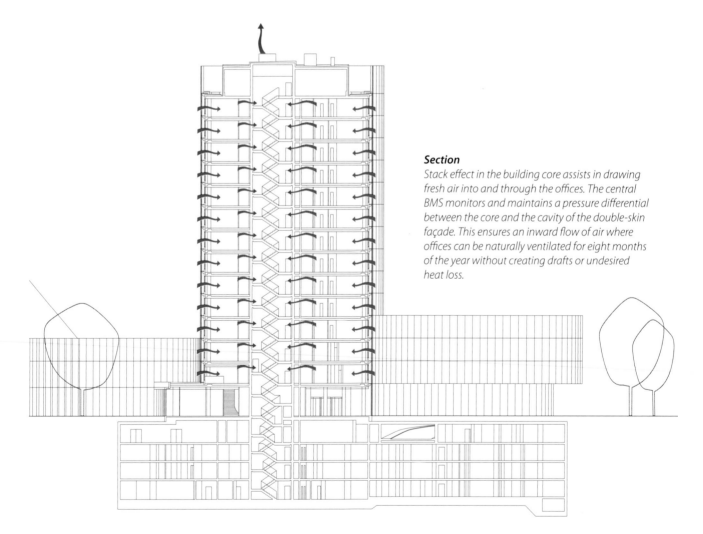

## Section

Stack effect in the building core assists in drawing fresh air into and through the offices. The central BMS monitors and maintains a pressure differential between the core and the cavity of the double-skin façade. This ensures an inward flow of air where offices can be naturally ventilated for eight months of the year without creating drafts or undesired heat loss.

▲ Figure 2.13.3: Typical office floor plan.
◀ Figure 2.13.4: Building section. (Base drawings © Sauerbruch Hutton)

▲ Figure 2.13.5: View of interior office space. © Jan Bitter

▲ Figure 2.13.6: Sawtooth profile façade with slender colored panels between larger fixed glass panels. © Jan Bitter

**The outer skin of the façade forms a sawtooth profile with fixed glass panels alternating with slender colored glass panels that open up to 90°, allowing fresh air into the cavity of the façade.**

The podium houses a reception foyer, a conference center, and cellular offices. The tower's central core houses vertical circulation, toilets, and service amenities. A corridor, which encircles the core, allows access to cellular offices arranged along the periphery of the building (see building plan and section, Figures 2.13.3 & 2.13.4). This arrangement allows all the offices to have access to daylight and natural ventilation (see Figure 2.13.5).

## Natural Ventilation Strategy

The cellular offices, located along the periphery of the building, are naturally ventilated via a double-skin façade which wraps the entire perimeter and is segmented at each floor. The outer skin of the façade forms a sawtooth profile alternating with fixed glass panels (the wider of the two) and slender colored glass panels (see Figure 2.13.6). The colored glass panels alternate between a fixed panel with acoustic insulation and an operable panel that opens up to 90° (side-hung), allowing fresh air into the cavity of the façade.

The inner skin also alternates between fixed and operable units consisting of argon-filled insulated windows with a low-E coating. The 700 mm cavity is horizontally continuous along the entire building perimeter (see Figures 2.13.7 & 2.13.8) and partitioned at each slab level with steel sheeting for fire protection. Fire protection sheets are also integrated in the vertical parapets on the inner side of the double-skin. Smoke protective curtains can sub-divide the cavity of the façade into three fire-protection zones which are analogous to the fire-protection zones of each floor. The sawtooth profile of

▲ Figure 2.13.7: View into the 700 mm continuous horizontal cavity of the double-skin. © KfW-Bildarchiv/Thomas Klewar

▲ Figure 2.13.8: Double-skin façade viewed from interior showing the colored glass panels that open by pivoting outwards. © Jan Bitter

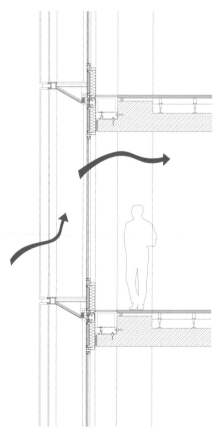

▲ Figure 2.13.9: Detailed section through the double-skin façade.
© Sauerbruch Hutton

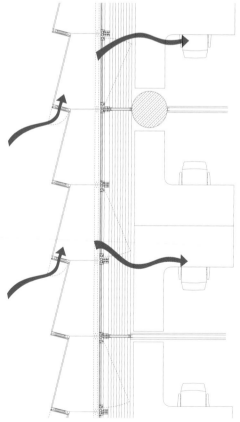

▲ Figure 2.13.10: Detailed plan through the double-skin façade.
© Sauerbruch Hutton

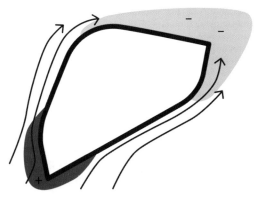
**Prevailing Winds**

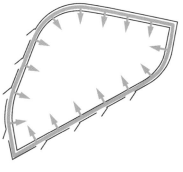
**Pressure Distribution**

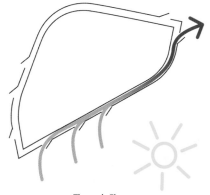
**Through-Flow**

▲ Figure 2.13.11: Diagrams of the Pressure Ring façade. © Sauerbruch Hutton

the façade's exterior and the absorptive aluminum sheeting (mounted above and below the steel sheeting which vertically segments the façade cavity), ensures that interior sound transmission is avoided between offices (see Figure 2.13.9).

The building is naturally ventilated when the outer colored panels are open, bringing fresh outdoor air into the offices through the operable windows on the inner side of the double-skin façade (see Figure 2.13.10). The panels are automatically opened depending on the external temperature and wind pressure, ensuring that the temperature and pressure conditions remain constant in the interstitial space. On warm summer days, the exterior façade can be opened on all sides, allowing wind to flow through the cavity preventing overheating. On cooler days, a minimal number of vents are open, enabling solar heating of the façade.

A unique feature of the double-skin façade is that the air cavity, also called the "Pressure Ring," maintains a ring of consistent positive pressure around the building (see Figure 2.13.11). The consistent positive pressure is made

available due to the orientation and operation of the colored flaps, which are monitored and operated by the building management system (BMS) (see "Interface with the Central Building Management System"). A common problem in naturally ventilated buildings with operable windows is pressure differences on the windward and leeward side, which can generate too much cross-ventilation, resulting in unwanted drafts and heat loss, particularly during the winter. The Pressure Ring in KfW Westarkade protects the tower against such undesirable effects of high wind speeds and ensures a constant regulated flow of air into the interior, allowing occupants to individually open the windows all year round without drafts or heat loss. As a result the ventilation of the offices is achieved naturally with minimal dependence on outside conditions.

As the air travels across the offices, it is subsequently exhausted to the negatively pressurized corridors encircling the offices, and ultimately via the building core through slots on the core walls just below the ceiling level, and then out through shafts which vent to the roof (see Figure 2.13.12). Once

**A unique feature of the double-skin façade is that the horizontally continuous air cavity, also called the Pressure Ring, maintains a ring of consistent positive pressure around the building.**

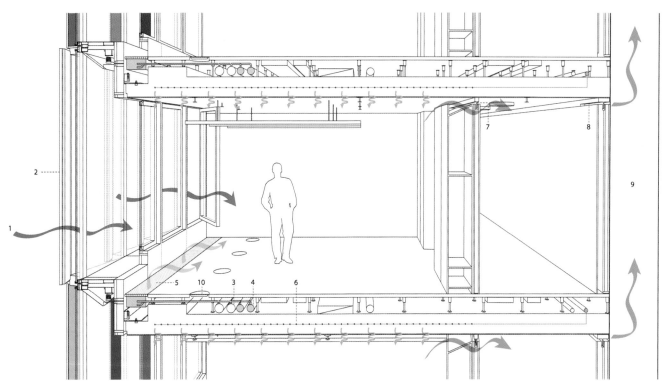

1 – Fresh air supply
2 – Air vent
3 – Central heating lines
4 – Central cooling lines
5 – Air vents or under-floor convectors in every second axis

6 – Thermo-active building components
7 – Sound-attenuating air overflow
8 – Central exhaust
9 – Vertical exhaust shaft using natural stratification
10 – Under-floor electrical outlet

▲ Figure 2.13.12: Detailed section of typical office ventilation. © Sauerbruch Hutton

**Air is pre-tempered through a 30-meter-long geothermal underground duct before it is supplied to the office spaces, thus optimizing the efficiency of mechanical ventilation within the building.**

the air reaches the cores, it is naturally driven upwards through stratification when wind speeds induce a sufficient negative pressure at the roof level. In the event of low wind speeds, mechanical ventilation is used to exhaust the air from the cores. During winter, when heat-recovery is implemented, mechanical ventilation is used and a by-pass flap is opened to bring warm exhaust air back to the HVAC center for recycling.

### Mixed-Mode Strategy

The building is Mixed-Mode as it is designed to operate under natural ventilation or mechanical ventilation

mode, depending on outdoor weather conditions. However, it is considered a Complementary-Concurrent system because users can open their windows at any time (though they are advised by LED displays in each office). Due to the designed efficacy of the Pressure Ring, the offices can be ventilated naturally for eight months of the year without creating drafts or undesired heat loss. This ensures that the use of mechanical air-conditioning is required for less than 50 percent of all working hours (Sauerbruch 2011).

Mechanically assisted air-conditioning is activated when temperatures within the double-skin façade are below 10 °C in the winter or above 25 °C in the summer (Meyer 2011). In this case, displacement ventilation is activated where fresh outdoor air, drawn through a duct buried beneath a below-grade parking garage, is supplied to the offices via a plenum below a raised floor system. The duct beneath the parking garage brings fresh outdoor air from an intake louver located at the edge of the site near the Palmengarten. The air is then pre-tempered through a 30-meter-long geothermal underground duct before it is supplied to the office spaces, thus optimizing the efficiency of mechanical ventilation within the building. In winter, the air will be further warmed by a heat-recovery system which uses both recycled thermal energy from exhaust air in the building and excess heat from the data processing center located in the basement (see Figure 2.13.13). Furthermore, additional heating or cooling of the air can be

achieved through the thermal activation of the radiant slabs, which feature integrated water pipes connected to a heat exchange system. During summer, the slabs maintain a constant 21 °C by being infused at night with water cooled to approximately 18 °C. Their concrete mass stores the coolness and then releases it during the day.

## Interface with the Central Building Management System

The building has a roof-mounted weather station that monitors the outdoor temperature, pressure, wind direction, wind speed, and daylight levels, among numerous other factors. Depending on the weather conditions, the BMS, opens or closes the ventilation flaps of the outer skin to introduce fresh air into the façade cavity and

subsequently into the building. The positioning of the ventilation flaps allows for numerous configurations. The flaps can be positioned anywhere from closed to fully open at 90° (see Figure 2.13.14). Which flaps, and how many, are opened usually depends on outside conditions as well as on the amount of windows opened on the inner façade; it is dynamically regulated to maintain an equal amount of pressure within the ring. The BMS controls the specific operation of the flaps in order to create a zone of consistent positive pressure around the building's inner skin, while simultaneously generating a slight pressure differential between the skin's cavity and the interior to maintain inward flow of air (Gonchar 2010). The airflow within the Pressure Ring is regulated and should not exceed six meters per second. Moreover, the operation of the flaps is also determined by the

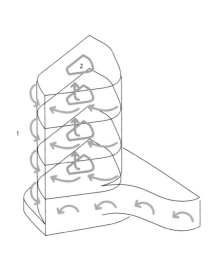

**Natural Ventilation Mode (Spring & Autumn)**

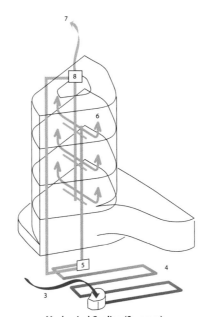

**Mechanical Cooling (Summer)**

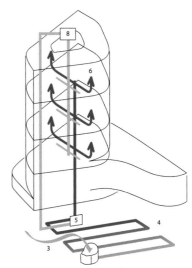

**Mechanical Heating (Winter)**

| | |
|---|---|
| 1 – Natural ventilation by manually operable windows | 5 – HVAC center |
| 2 – Mechanical ventilation of cores | 6 – Incoming air through raised floor |
| 3 – Incoming air from Palmengarten | 7 – Exhaust air shaft using stratification |
| 4 – Geothermal heat exchanger | 8 – Heat-recovery |

▲ Figure 2.13.13: Ventilation strategies, including mechanical ventilation modes when outdoor conditions inhibit natural ventilation. © Sauerbruch Hutton

▲ Figure 2.13.14: Façade detail showing condition where some colored flaps are opened to a 90° angle letting air into the façade cavity, while others remain closed (and thus, not visible in this image). © Jan Bitter

▲ Figure 2.13.15: Opening of the entire façade at the "tip" to enable through-flow of air on hot summer days. © Jan Bitter

**To prevent overheating on hot summer days, the entire façade (including the wider window panels which are typically fixed) at the "tip" of the building can be opened to enable through-flow of air.**

outdoor temperature. They can be fully opened in the summer to avoid overheating, and minimally opened in the winter to allow the double-skin façade to function as a solar collector. To prevent overheating on hot summer days, the entire façade (including the wider window panels which are typically fixed) at the "tip" of the building can be opened to enable through-flow of air (see Figure 2.13.15). Throughout the entire year, the BMS will advise occupants whether or not to open their windows through an LED panel in offices, but it gives the occupant the final choice.

### Other Sustainable Design Elements

A number of tightly coordinated sustainable design strategies were employed to help the building meet its ambitious target of employing less than 100 kWh/m² of primary energy

annually. These strategies include the use of radiant slabs with a geothermal heat exchange system, high thermal insulation levels, an efficient heat-recovery system, and a ground air exchange system. In addition, the cavity of the double-skin façade incorporates automated venetian blinds with a light redirection feature which controls glare, minimizes solar heat gain, and reflects daylight deep into the building. The operation of the blinds is also controlled by the BMS.

### Analysis – Performance

From the planning outset, ambitious goals were set to comply with guidelines for administrative buildings set forth by the incentive program "SolarBau." SolarBau stipulated a maximum primary energy usage of 100 kWh/m² (total building) and internal building temperatures should remain below 26 °C. Based on simulations, energy consumption could be as low as 82 kWh/m$^2$. It is also predicted that the annual carbon footprint would be 43 kg $CO_2$/m$^2$. The project is currently undergoing a rigorous monitoring program by researchers from the University of Karlsruhe.

## Analysis – Strengths

▶ The aerodynamic wing-shaped form of the tower and its orientation along the prevailing wind direction allow for minimization of pressure build up around the building. Further, the design of the double-skin façade with its mechanically operable ventilation flaps (inlets) maintain a uniform pressure in the cavity. This dynamic system negates the effects of variable pressure around the tower (a common problem in tall building design), reducing unwanted drafts, heat loss, and enabling natural ventilation for much of the year.

▶ The maintenance and uniformity of positive pressure in the cavity of the double-skin façade enables a consistent, controlled, and regulated flow of pre-tempered air into the building, thus enhancing the effectiveness of natural ventilation.

▶ The optimized operability of outer ventilation flaps (as they can be fully opened) minimizes the risk of overheating within the double-skin façade.

▶ Prefabricated modules of the saw-tooth, double-skin façade enabled rapid assembly, independent of weather changes, and guaranteed exceptionally high construction quality.

## Analysis – Considerations

The following areas could be causes for concern if adopting similar strategies in other buildings and should therefore be considered:

▶ The BMS system must be carefully designed to precisely control the façade openings in order to maintain the Pressure Ring in all wind conditions (i.e., not just from the prevailing direction).

▶ If the interior arrangement is such that offices at different façades belong to the same supply air group, the temperature in both façades must fall within an acceptable range for natural ventilation. Due to different solar irradiation on each façade, this may significantly reduce the amount of time these offices can be naturally ventilated.

▶ The feasibility of a Pressure Ring system may reduce as the height of the building increases (especially in taller buildings) due to the wind velocities/pressures at height becoming too great.

## Project Team

**Owner/Developer:** KfW Bankengruppe
**Design Architect:** Sauerbruch Hutton
**Associate Architect:** Architekten Theiss Planungsgesellschaft mbH
**Structural Engineer:** Werner Sobek Group
**Energy Concept:** Transsolar
**MEP Engineer:** Reuter Rührgartner GmbH; Zibell, Willner & Partner
**Main Contractor:** ARGE Züblin; Bögl

**References & Further Reading**

**Books:**

▶ Wood, A. (ed.) (2011) *Best Tall Buildings 2011: CTBUH International Award Winning Projects.* Routledge: New York, pp. 118–123.

**Journal Articles:**

▶ Gonchar, J. (2010) "More than skin deep," *Architectural Record,* vol. 198, no. 7, pp. 102–110.

▶ Meyer, U. (2011) "Colors 'n curves: a new bank headquarters in Frankfurt may well be the world's most energy-efficient office tower – KFW Westarkade, Frankfurt, Germany," *GreenSource Magazine,* pp. 48–53.

▶ Sauerbruch, M. (2011) "Sustainable architecture," *Detail Green English,* vol. 1, pp. 26–31.

## Project Data:

**Year of Completion**
▸ 2011

**Height**
▸ 139 meters

**Stories**
▸ 30

**Gross Area of Tower**
▸ 55,000 square meters

**Building Function**
▸ Office

**Structural Material**
▸ Concrete

**Plan Depth**
▸ 23.5 meters (from void)

**Location of Plant Floors:**
▸ B1–B3, 17, 30

## Ventilation Overview:

**Ventilation Type**
▸ Mixed-Mode: Zoned

**Natural Ventilation Strategy**
▸ Cross and Stack Ventilation (only in lobby, atrium, and break-out areas)
▸ Office spaces can be upgraded to natural ventilation

**Design Strategies**
▸ Naturally ventilated atrium, lobby, and break-out areas

**Double-Skin Façade Cavity:**
▸ Depth: 600 mm
▸ Horizontal Continuity: Fully Continuous (around entire perimeter)
▸ Vertical Continuity: 3.85 meters (floor-to-floor)

**Approximate Percentage of Year Natural Ventilation can be Utilized:**
▸ 100% (only in lobby, atrium, and break-out areas)

**Percentage of Annual Energy Savings for Heating and Cooling:**
▸ 63% compared to a fully air-conditioned Australian office building (estimated)

**Typical Annual Energy Consumption (Heating/Cooling):**
▸ Unpublished

# Case Study 2.14
# 1 Bligh Street Sydney, Australia

## Climate

Sydney has a temperate climate with warm summers and mild winters, moderated by proximity to the ocean. The summer months of December, January, and February are known for high temperatures averaging around 26 °C. However, cloud-free skies and extremely sunny days can create summer temperatures which approach 40 °C. Rainfall is evenly spread throughout the year, August and September being the driest months. Winter weather remains warm with an average temperature of 17 °C (see Figure 2.14.1).

## Background

1 Bligh Street aims to provide an energy efficient environment with access to daylight and ventilation, and large flexible floor plates. The building is prominently located in Sydney's Central Business District. In response to a site skewed at a 45-degree angle from the downtown street grid, a rotated elliptical form was used (see Figure 2.14.2). The long axis of the ellipse runs along the southwest and northeast direction, optimizing views to the harbor. The first floor accommodates communal spaces such as an outdoor covered café, a child care center, and a restaurant and bar. Most of these first-floor facilities are open outdoor spaces protected by the overhang of the floors above.

The views from the office floors are maximized by placing the circulation core on the south side of the tower (see plan and section, Figures 2.14.3 & 2.14.4) adjacent to surrounding buildings. A naturally ventilated atrium is the communal "heart" of the tower, which enhances visual communication and maximizes access to daylight and fresh air. The inner walls of the atrium are flanked by glass lifts, corridors, meeting rooms, and terraced balconies (see Figure 2.14.5). The office floors are open-plan with the option of adding cellular offices along the perimeter. Two sky gardens are located in the tower, one on level 15 adjacent to the elevator transfer lobby and the other located on the rooftop shielded by the outer skin of the façade, which rises ten meters above the topmost floor.

## Climatic Data:[1]

**Location**
▸ Sydney, Australia

**Geographic Position**
▸ Latitude 34° 0'S, Longitude 151° 0'E

**Climate Classification**
▸ Tropical/Temperate (mild seasonal variation)

**Prevailing Wind Direction**
▸ North-northeast

**Average Wind Speed**
▸ 3.8 meters per second

**Mean Annual Temperature**
▸ 18 °C

**Average Daytime Temperature during the Hottest Months (December, January, February)**
▸ 26 °C

**Average Daytime Temperature during the Coldest Months (June, July, August)**
▸ 17 °C

**Day/Night Temperature Difference During the Hottest Months**
▸ 7 °C

**Mean Annual Precipitation**
▸ 1,222 millimeters

**Average Relative Humidity**
▸ 66% (hottest months); 62% (coldest months)

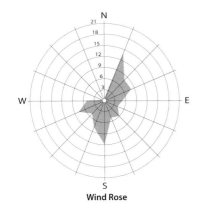

**Wind Rose**

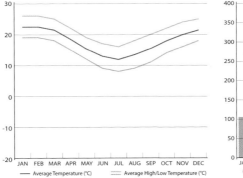

**Average Annual Temperature Profile (°C)**
— Average Temperature (°C)  ---- Average High/Low Temperature (°C)

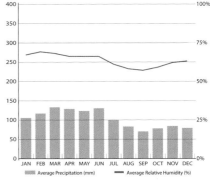

**Average Relative Humidity (%) and Average Annual Rainfall**
▢ Average Precipitation (mm)  — Average Relative Humidity (%)

▲ Figure 2.14.1: Climate profiles for Sydney, Australia.[1]
◂ Figure 2.14.2: Overall view. © DEXUS Property Group

---

[1] The climatic data listed for Sydney was derived from the World Meteorological Organization (WMO) and the Bureau of Meteorolgy, Australia.

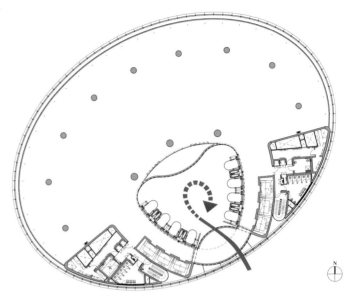

### Plan

*The lobby, central atrium, and adjacent corridors and meeting spaces are naturally ventilated through stack effect. Fresh air enters the building through pivoting glass panels in the lobby's curtain wall and mid-height at the sky garden level. An opening at the top of the atrium exhausts the air. Office spaces are fully mechanically ventilated.*

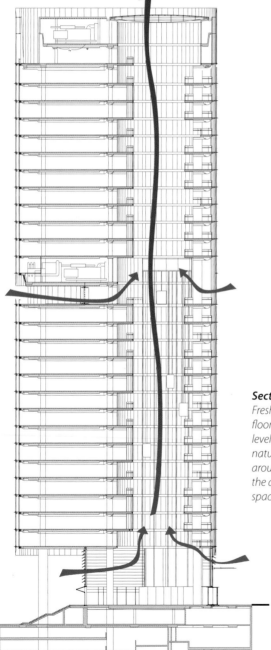

### Section

*Fresh air is brought in through glass louvers within the ground floor lobby's exterior wall and mid-height at the sky garden level. Stack effect draws the fresh air through the atrium, naturally ventilating the balconies and circulation spaces around the atrium. Treated air from the offices also flows into the atrium, providing tempered air to further condition the space, providing comfortable conditions for much of the year.*

▲ Figure 2.14.3: Typical office floor plan.
◄ Figure 2.14.4: Building section. (Base drawings © ingenhoven architects)

▲ Figure 2.14.5: View looking up through the naturally ventilated atrium. © ingenhoven architects

**The naturally ventilated atrium is the communal "heart" of the tower and naturally ventilates the balconies and corridors which surround it on each floor.**

## Natural Ventilation Strategy

1 Bligh Street has a 120-meter-high naturally ventilated atrium forming the social spine of the building. Fresh air flows into the atrium through the sky garden at mid-height and operable glass louvers in the façade of the ground floor lobby (see Figure 2.14.6). Stack effect exhausts the atrium through openings in the glass roof of the atrium. The atrium naturally ventilates balconies and corridors which surround it on each floor.

The office areas of the building are fully mechanically ventilated and separated from the atrium through glass walls and doors. To meet Australia's Green Star energy ratings, a high turnover of air in the office spaces was required, but rather than constantly exhaust large amounts of conditioned air, the building allows the "used air" of the offices to flow into the atrium, providing tempered air to cool the occupied atrium areas in summer and heat them in winter, providing comfortable conditions for much of the year.

The building's exterior has a double-skin façade consisting of a double-glazed insulating layer with a very high-performance low-E coating and a single-glazed laminated low-iron glass.

The 600 mm cavity of the façade is horizontally continuous along the length of the façade, but vertically segmented at each slab level. The outer glass panels of the double-skin façade were engineered with fixed louvers at the edge of each floor slab, allowing fresh air to enter at the bottom of each cavity and exhaust at the top (see Figure 2.14.7).

The aerofoil-shaped louvers were designed using computational fluid dynamics and wind-tunnel tests to ensure high wind speeds would not affect the blinds contained within, and exhaust air would not re-enter the cavity above. Airflow within the façade cavity helps maintain a constant average temperature in the building, reducing the reliance on the HVAC system (Seguin 2011). However, the double-skin façade is not used to naturally ventilate the building, thus the inner façade is not operable, and is solely used to protect the blinds.

The tower was designed to potentially allow natural ventilation from the perimeter if the tenants chose to upgrade and retrofit the interior glass panels with a hardware allowing them to be operable (Meyer 2008). But even in this scenario, the floor plates are too deep for natural ventilation across

them, therefore natural ventilation in the offices would likely be limited to just perimeter cellular offices.

## Mixed-Mode Strategy

The building can be considered a Mixed-Mode: Zoned building with mechanical cooling and natural ventilation operating in different areas of the building, mostly independently. The lobby, atrium, balconies, and corridors are naturally ventilated throughout the year. The office spaces are conditioned in two zones, using highly efficient chilled beams at the perimeter, and a variable-air-volume system in the interior. The perimeter zone has its own dedicated air-handling units so they could be shut down in isolation should a future retrofit allow the perimeter zone to be naturally ventilated for parts of the year. The lobby floor incorporates in-slab heating which uses waste heat from the HVAC system to provide additional warmth in the winter and temper the air in the atrium.

## Interface with the Central Building Management System

The building management system (BMS) has full control of the natural ventilation in the atrium. It is linked to

▲ Figure 2.14.6: Exterior view of lobby showing operable glass louvers.
© ingenhoven architects

▲ Figure 2.14.7: View of aerofoil louver at the edge of each floor slab that allow fresh air into the double-skin cavity. © ingenhoven architects

the weather stations and numerous sensors which measure temperature, humidity, wind speeds, rainfall, and light levels. Should the atrium become too hot while outdoor conditions are cool, the atrium will open completely. If the outdoor temperatures are too hot, the atrium will be sealed off and tempered by the pre-conditioned spill air from the offices alone. If wind conditions in the atrium are too high, or there is the potential for wind-driven rain, the louvers will be closed off. The BMS also controls the operation of the blinds in the cavity of the double-skin façade using sun-tracking software and photo-sensors that enable the blinds to respond to the variations of the sun's angle and external light levels.

## Other Sustainable Design Elements

As mentioned previously, the building has automatic, mechanically operated venetian blinds to protect the building from solar glare and heat gain. By placing the blinds within the double-façade cavity they are more effective at reducing solar heat gain before it enters the building. The blinds also allow for the use of clear glass for optimal daylight penetration (see Figure 2.14.8).

The building is equipped with a hybrid tri-generation system that simultaneously generates heating,

cooling, and electrical power. A gas-fired cogeneration generates electricity and useful heat, whereas roof-mounted solar thermal collectors generate energy (which feeds into a tri-generation system) to power an absorption chiller that drives the cooling systems.

## Analysis – Performance

Due to the recent completion of this building at the time of this publication, there is little performance information available. The building did achieve a 6-Star Green Star Design rating, and at the time of publication was under assessment for a 6-Star Green Star Office As-Built rating. The building is also targeting a 5-Star NABERS (National Australian Built Environment Rating System) rating, though this rating can only be achieved after the building has been in operation for 12 months of occupancy (Linn 2011). The energy savings for the heating and cooling of the building are estimated to be approximately 63 percent over a typical Australian office building, however, the actual energy performance of the building is the result of a combination of many strategies, with the natural ventilation of the atrium only one part.

Considering Sydney's temperature and humidity conditions and then assessing the impact of internal heat loads, it has

been estimated that natural ventilation could be suitable for approximately 35–40 percent of the year in the office spaces (after retrofit). This can further reduce the building's reliance on air-conditioning and potentially result in additional energy savings.

## Analysis – Strengths

▸ The compact elliptical form increases the ratio of the building's volume-to-surface area, therefore reducing heat gain and loss through the envelope and optimizing energy performance. The form also minimizes wind turbulence and downdrafts, improving the environment at street level.

▸ Compared to a conventional office building with a central circulation core that requires artificial lighting and conditioning, the atrium is exploited as a day-lit and naturally ventilated circulation space, minimizing electricity demands and HVAC energy consumption.

▸ The atrium forms the social hub of the building, and balconies that project into the atrium provide naturally ventilated break-out spaces and create opportunities for social interaction while enhancing visual connectivity.

- The exhausted, conditioned air of the office spaces is "re-used" to temper the naturally ventilated atrium year-round.

- The external skin is effectively used to protect the blinds from direct wind, allowing for a fully glazed, transparent building with a high visual light transmission that both mitigates heat gain and meets the high-performance targets.

- The aerofoil-shaped louvers direct wind into the cavity without damaging the blinds, and help avoid short-circuiting problems of intake and exhaust air between the separated double-skin cavities.

- A "zoned" mixed-mode ventilation strategy (where occupied office spaces are air-conditioned and public/circulation zones are naturally ventilated) could be very suitable for tall buildings in places that experience hot and humid climates. Occupants tend to have less temperature demands in a circulation space as opposed to a working space.

would allow for single-sided ventilation along the perimeter zones; however, the plan depths are too deep to effectively ventilate the entire office floor.

- The building cores could have been used as solar buffers; however, the orientation and spatial configuration of the building are mainly derived from the urban conditions and the desire to maximize views, not by environmental concerns.

## Project Team

**Owner/Developer:** DEXUS Property Group; Cbus Property; Dexus Wholesale Property Fund
**Design Architect:** ingenhoven architects
**Structural Engineer:** Enstruct Group
**MEP Engineer:** Arup
**Environmental Consultant:** Cundall
**Project Manager:** APP Corporation
**Main Contractor:** Grocon
**Other Consultants:** DS-Plan AG & Arup (Façade Consultant)

**References & Further Reading**

**Journal Articles:**

- Lehmann, S. & Ingenhoven, C. (2009) "The future is green: a conversation between two German architects in Sydney," *Journal of Green Building*, vol. 4, no. 3, pp. 44–51.

- Linn, C. (2011) "Architects: Architectus and Ingenhoven Architects," *Architect*, vol. 100, no. 11, pp. 84–88.

- Meyer, U. (2008) "Double-skin deep: designing environmentally sustainable architecture often depending on the façade," *World Architecture (China)*, vol. 214, pp. 20–25.

- Seguin, B. (2011) "Façade technology," *Architecture Australia*, vol. 100, no. 3, pp. 104–105.

- Vivian, P. (2008) "Space: next-generation workspace," *Architecture Australia*, vol. 97, no. 3, pp. 97–102.

## Analysis – Considerations

The following areas could be causes for concern if adopting similar strategies in other buildings and should therefore be considered:

- In a relatively mild climate that could allow natural ventilation for most of the year, the use of fixed, non-operable windows on a double-skin façade for the office spaces is a lost opportunity for natural ventilation of the entire space.

- In the case of a façade retrofit, opening the internal windows

▲ Figure 2.14.8: Internal view of double-skin façade. © DEXUS Property Group

# 3.0 Design Considerations, Risks, and Limitations

# 3.0 Design Considerations, Risks, and Limitations

The table on the following pages (142–143) shows comparative data summarizing the 14 case studies. This section of the guide discusses common issues that have arisen out of the case studies and should be considered in the natural ventilation of a tall office building. Although this guide has concentrated on the office typology (since office space is difficult to naturally ventilate due to high-density occupancy, high internal heat gains, and low common tolerance among the building population to temperature variations), the issues discussed here are also relevant to other building types, including residential, hotel, mixed-use, etc.

## 3.1 Thermal Comfort Standards

The main purpose of ventilating buildings (whether naturally or mechanically) is to provide a comfortable internal environment and acceptable air quality. It is therefore important to understand the wider issues of thermal comfort standards before considering detailed aspects of natural ventilation strategies.

Historically the concept of thermal comfort was defined by Fanger, using the concept of Predicted Mean Vote (PMV). The PMV index predicts thermal sensation on a seven point scale where −3 represents very cold, 0 represents neutral, and +3 represents very hot. The American Society of Heating, Refrigeration, and Air-Conditioning Engineers developed ASHRAE Standard 55, which was based on the PMV model, to determine thermal comfort in mechanically conditioned spaces.

ASHRAE Standard 55 contains, arguably, a narrow definition of thermal comfort which cannot be easily achieved in naturally ventilated buildings, even in relatively mild climatic zones (Brager & de Dear 2000). In 2004, ASHRAE Standard 55 issued an additional optional method for determining acceptable thermal conditions in naturally conditioned spaces (see Figure 3.1). However, the code still stipulates that if the space is mixed-mode and includes any mechanical cooling strategies, then the building must still adhere to the strict standard for conditioned spaces.

It is important to understand that the differences in thermal comfort standards used for naturally ventilated buildings and air-conditioned buildings can be influenced significantly by context. People living or working year-round in air-conditioned spaces tend to develop high expectations for homogeneity of air temperature and humidity, with little tolerance of variations. In contrast, people who live and/or work in naturally ventilated buildings adapt to more variable indoor thermal comfort conditions that reflect outdoor local climate variations (Brager & de Dear 2002). Inherently, their preference and tolerance are likely to extend over a wider temperature range than that stated by ASHRAE Standard 55 (see Figure 3.2).

Further, this tolerance to variations in internal environment and comfort is influenced by a wide range of cultural, socio-economic, and climatic factors. Acceptable ranges for internal air temperature and humidity will vary significantly across countries, cultures, age groups, and numerous other factors. When designing a naturally

## When designing a naturally ventilated building it is beneficial to adopt "adaptive" thermal comfort standards which takes cultural, social, and contextual factors into account.

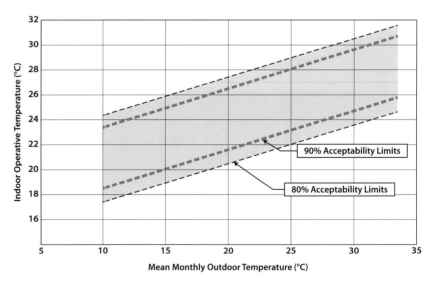

▲ Figure 3.1: Adaptive standards for naturally ventilated buildings. (Source: ASHRAE Standard 55-2004)

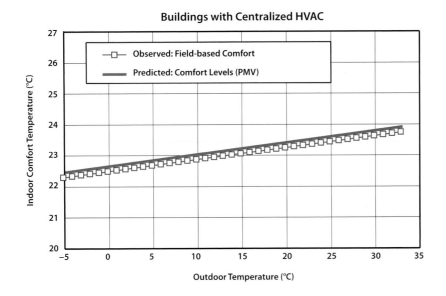

**Buildings with Centralized HVAC**

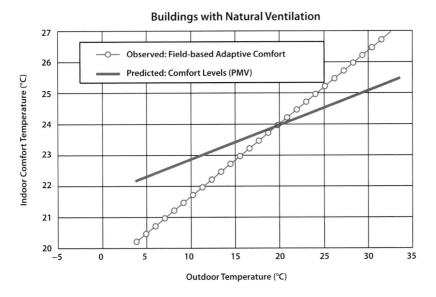

**Buildings with Natural Ventilation**

▲ Figure 3.2: Comfort temperatures based on adaptive versus predicate mean vote (PMV) models for buildings with a centralized HVAC system (top) and buildings with natural ventilation (bottom). (Source: Brager, G. S. & de Dear, R. (2000) p. 26)

ventilated building, it is beneficial to adopt an "adaptive" thermal comfort standard which takes these cultural, social, climatic, and contextual factors into account. In addition, when building occupants have control over their environment, thermal comfort ranges can be further extended beyond the normal range. As in the example of the Commerzbank, Frankfurt (see pages

32–41), workers in offices facing the internal atrium preferred to keep the windows open and receive natural ventilation from the atrium, at times accepting a higher air temperature, rather than switch to mechanical ventilation. This has developed to the point where these zones of the building are now 100 percent naturally ventilated the entire year round, irrespective

of mechanical ventilation being used during peak season in other parts of the building. By following "adaptive" standards and allowing occupants direct control over their environment, one can extend the range of thermal comfort conditions and enhance the prospects of a naturally ventilated tall building.

Certainly more post-occupancy research is needed before we fully understand the range of acceptable internal environmental conditions to which we can design our buildings. Ultimately a far wider range of internal environment variations may need to be accepted (including clothing attire, working patterns, and acceptance of occasional "uncomfortable" peak conditions) to truly benefit from the reduced energy, cost, and space savings inherent in a fully naturally ventilated building. It may be that the advances in the technical field of natural ventilation need to also be matched with changes in attitude among building occupants. This latter aspect may prove to be more challenging than the technical or design challenges.

## 3.2 Local Climate

Climate considerations are a key aspect to any design that aims to be energy efficient and provide environmental comfort largely by natural means. Natural ventilation strategies can be derived according to general climate classifications (of which there are four: tropical, hot dry, temperate, and cold), as well as detailed investigation of the impact of the climate on the site directly. As different climates pose different environmental requirements, various design strategies can be generated to achieve human comfort. Such strategies can lead to significant energy savings and therefore a better energy performance whether the building is

# Case Study Comparative Analysis Table

| | RWE Headquarters Tower, Essen, 1996 | Commerzbank, Frankfurt, 1997 | Liberty Tower of Meiji University, Tokyo, 1998 | Menara UMNO, Penang, 1998 | Deutsche Messe AG Building, Hannover, 1999 | GSW Headquarters Tower, Berlin, 1999 | Post Tower, Bonn, 2002 |
|---|---|---|---|---|---|---|---|
| Climate Classification | Temperate | Temperate | Temperate | Tropical | Temperate | Temperate | Temperate |
| Mean Annual Temperature | 10 °C | 10 °C | 17 °C | 28 °C | 9 °C | 9 °C | 10 °C |
| Average Daytime Temp. – Hottest | 22 °C | 24 °C | 29 °C | 32 °C | 22 °C | 23 °C | 17 °C |
| Average Daytime Temp. – Coldest | 5 °C | 4 °C | 11 °C | 31 °C | 5 °C | 4 °C | 3 °C |
| Day/Night Temp. Difference | 9 °C | 11 °C | 7 °C | 8 °C | 10 °C | 10 °C | 11 °C |
| Average Relative Humidity | 73% (hottest months) 82% (coldest months) | 53% (hottest months) 76% (coldest months) | 72% (hottest months) 50% (coldest months) | 79% (hottest months) 73% (coldest months) | 72% (hottest months) 85% (coldest months) | 70% (hottest months) 85% (coldest months) | 72% (hottest months) 83% (coldest months) |
| Wind Speed (direction) | 2.7 m/s (west-southwest) | 4 m/s (south-southwest) | 3.4 m/s (south) | 2.6 m/s (south-southwest) | 2.6 m/s (west) | 2.7 m/s (west) | 2.4 m/s (west-northwest) |
| Occupier | Owner | Owner | Owner | Multi-tenant | Owner | Multi-tenant | Owner |
| Height (number of floors) | 127 meters (31 floors) | 259 meters (56 floors) | 119 meters (23 floors) | 94 meters (21 floors) | 82 meters (20 floors) | 82 meters (23 floors) | 163 meters (42 floors) |
| Plan Depth | 8 meters (from central core) | 16.5 meters (from central void) | 20 meters (from core) | 14 meters (from core) | 24 meters (between façades) | 7.2–11 meters (between façades) | 12 meters (from central void) |
| Ventilation Type | Mixed-Mode: Complementary-Changeover | Mixed-Mode: Complementary-Changeover | Mixed-Mode: Complementary-Changeover | Mixed-Mode: Complementary-Alternate | Mixed-Mode: Complementary-Changeover | Mixed-Mode: Complementary-Changeover | Mixed-Mode: Zoned / Complementary-Changeover |
| Natural Ventilation Driving Force | ▸ Wind-Driven Single-Sided Ventilation | ▸ Cross and Stack Ventilation (connected internal spaces) | ▸ Cross and Stack Ventilation (connected internal spaces) | ▸ Wind-Driven Cross-Ventilation | ▸ Wind-Driven Cross-Ventilation ▸ Stack Ventilation via Ventilation Tower | ▸ Single-Sided & Cross-Ventilation ▸ Stack Ventilation via External Thermal Flue | ▸ Cross and Stack Ventilation (connected internal spaces) |
| Design Strategies | ▸ "Fish Mouth" device which adjusts air intake speed ▸ Aerodynamic external form | ▸ Stepping sky gardens connected by segmented central atrium ▸ Small aerofoil sections above/below ventilation slots | ▸ Ventilation "Wind Core" (central escalator void) ▸ "Wind Floor" over central void ▸ Innovative window openings | ▸ "Wing Walls" which capture a wider range of wind directions | ▸ "Corridor" double-skin façade (large-volume horizontal air duct) ▸ Ventilation tower to exhaust building | ▸ Double-skin as external thermal flue ▸ "Wing Roof" accelerates wind passing over the thermal flue which creates negative pressure | ▸ Full-height central atrium divided into 9-story sky gardens ▸ "Wing Wall" extensions ▸ Aerodynamic external form |
| Double-Skin Façade (see Figure 3.11 for more detail) | Yes; Cavity Depth: 500 mm | Yes; Cavity Depth: 200 mm | None | None | Yes; Cavity Depth: 1,400 mm | Yes; Cavity Depth: 1,000 mm (west façade); 200 mm (east façade) | Yes; Cavity Depth: 1,700 mm (south façade); 1,200 mm (north façade) |
| Use of Atrium or Sky Garden | None | Segmented Central Atrium and Stepped 4-Story Sky Gardens | Central Escalator Core | None | None | None | Three 9-Story Sky Gardens and one 11-Story Sky Garden |
| Control of Openings | Occupant Controlled | Automatically Controlled and Occupant Controlled | Automatically Controlled | Occupant Controlled | Occupant Controlled | Automatically Controlled and Occupant Controlled | Automatically Controlled and Occupant Controlled |
| Night-Time Ventilation | Yes | Yes | Yes | None | None | Yes | Yes |
| Approx. Percent of Year Natural Vent. can be Utilized | 75% | 80% | 29% | 0–100% (depending on tenant) | Unpublished | 70% | Unpublished |
| Percentage of Annual Energy Savings for Heating and Cooling | Unpublished | 63% compared to a fully air-conditioned German office building (measured) | 55% compared to a fully air-conditioned office building in Japan (measured) | 25% compared to a fully air-conditioned office building in Malaysia (measured) | Unpublished | 53% compared to a fully air-conditioned German office building (estimated) | 79% compared to a fully air-conditioned German office building (measured) |
| Typical Annual Energy Consumption (Heating/Cooling) | Unpublished | 117 kWh/m² (measured) | 166 kWh/m² (measured) | 180 kWh/m² (measured) | 43 kWh/m² (heating only) (estimated) | 150 kWh/m² (estimated) | 75 kWh/m² (measured) |

| 30 St. Mary Axe, London, 2004 | Highlight Towers, Munich, 2004 | Torre Cube, Guadalajara, 2005 | San Francisco Federal Building, 2007 | Manitoba Hydro Place, Winnipeg, 2008 | KfW Westarkade, Frankfurt, 2010 | 1 Bligh Street, Sydney, 2011 | |
|---|---|---|---|---|---|---|---|
| Temperate | Temperate | Temperate (mild seasonal variation) | Temperate (mild seasonal variation) | Cold | Temperate | Tropical/Temperate (mild seasonal variation) | **Climate Classification** |
| 11 °C | 9 °C | 20 °C | 14 °C | 3 °C | 10 °C | 18 °C | **Mean Annual Temperature** |
| 22 °C | 22 °C | 32 °C | 23 °C | 25 °C | 24 °C | 26 °C | **Average Daytime Temp. – Hottest** |
| 8 °C | 4 °C | 25 °C | 14 °C | −10 °C | 4 °C | 17 °C | **Average Daytime Temp. – Coldest** |
| 9 °C | 10 °C | 19 °C | 10 °C | 13 °C | 11 °C | 7 °C | **Day/Night Temp. Difference** |
| 66% (hottest months) 81% (coldest months) | 72% (hottest months) 84% (coldest months) | 52% (hottest months) 60% (coldest months) | 69% (hottest months) 69% (coldest months) | 69% (hottest months) 77% (coldest months) | 53% (hottest months) 76% (coldest months) | 66% (hottest months) 62% (coldest months) | **Average Relative Humidity** |
| 3.6 m/s (southwest) | 2.3 m/s (west) | 4.8 m/s (west) | 4.3 m/s (west) | 4.7 m/s (south) | 4 m/s (south-southwest) | 3.8 m/s (north-northeast) | **Wind Speed (direction)** |
| Multi-tenant | Multi-tenant | Multi-tenant | Owner | Owner | Owner | Multi-tenant | **Occupier** |
| 180 meters (42 floors) | 126 meters (33 floors) | 60 meters (17 floors) | 71 meters (18 floors) | 115 meters (22 floors) | 56 meters (14 floors) | 139 meters (30 floors) | **Height (number of floors)** |
| 6.4–13.1 meters (from central core) | 13.5 meters (between façades) | 9–12 meters (from central void) | 19 meters (between façades) | 11.5 meters (from central core) | 6.3 meters (from central core) | 23.5 meters (from void) | **Plan Depth** |
| Mixed-Mode: Complementary-Concurrent | Mixed-Mode: Complementary-Concurrent | Natural Ventilation (no mechanical) | Mixed-Mode: Zoned / Complementary-Concurrent | Mixed-Mode: Complementary-Concurrent | Mixed-Mode: Complementary-Concurrent | Mixed-Mode: Zoned | **Ventilation Type** |
| ▸ Cross and Stack Ventilation (connected internal spaces) | ▸ Wind-Driven Cross-Ventilation ▸ Stack Ventilation via Vertical Shaft | ▸ Cross and Stack Ventilation (connected internal spaces) | ▸ Wind-Driven Cross-Ventilation | ▸ Cross and Stack Ventilation (connected internal spaces) | ▸ Wind-Driven Cross-Ventilation ▸ Stack Ventilation via Vertical Shaft | ▸ Cross and Stack Ventilation (only in lobby, atrium, and break-out areas) | **Natural Ventilation Driving Force** |
| ▸ Stepping atria which tempers air before being distributed to offices | ▸ High-performance single-skin façade with perforated panels ▸ Narrow plan depth | ▸ Rain screen/brise-soleil façade ▸ Central (open) atrium ▸ Funnel-shaped office spaces | ▸ Thermal mass ▸ Night cooling ▸ Voids between cellular offices and ceilings for cross airflow | ▸ Segmented atria/sky gardens act as a thermal buffer zone ▸ 115-meter solar chimney | ▸ "Pressure Ring" which maintains a ring of consistent positive pressure ▸ Aerodynamic external form | ▸ Naturally ventilated atrium, lobby, and break-out areas | **Design Strategies** |
| Yes; Cavity Depth: 1,000–1,400 mm | None | None | None | Yes; Cavity Depth: 1,300 mm | Yes; Cavity Depth: 700 mm | Yes; Cavity Depth: 600 mm | **Double-Skin Façade** (see Figure 3.11 for more detail) |
| 2-Story and 6-Story Stepped, Spiraling Sky Gardens | None | Central Atrium and Stepped 4-Story Sky Gardens | None | 3-Story and 6-Story Stacked Sky Gardens | None | Central Atrium | **Use of Atrium or Sky Garden** |
| Automatically Controlled and Occupant Controlled | Automatically Controlled and Occupant Controlled | Occupant Controlled | Occupant Controlled | Automatically Controlled | Occupant Controlled | Not Applicable | **Control of Openings** |
| None | None | None | Yes | None | None | None | **Night-Time Ventilation** |
| 40% (as originally designed) | Unpublished | 100% | 75% | 35% | 60% | 100% (only in lobby, atrium, and break-out areas) | **Approx. Percent of Year Natural Vent. can be Utilized** |
| Unpublished | 69% compared to a fully air-conditioned German office building (estimated) | 100% compared to a conventional air-conditioned building of the same size and typology (assumed) | 55% compared to a conventional air-conditioned building of the same size and typology (estimated) | 73% compared to a fully air-conditioned office building in Manitoba (measured) | 84% compared to a fully air-conditioned German office building (estimated) | 63% compared to a fully air-conditioned office building in Australia (estimated) | **Percentage of Annual Energy Savings for Heating and Cooling** |
| Unpublished | 100 kWh/m² (estimated) | 0 kWh/m² (assumed) | Unpublished | 39 kWh/m² (measured) | 50 kWh/m² (estimated) | Unpublished | **Typical Annual Energy Consumption (Heating/Cooling)** |

▲ Figure 3.3: In a tropical climate, a fully naturally ventilated building may not be feasible, in which case a "Zoned" approach with separate consideration of mechanically ventilated office spaces and naturally ventilated circulation spaces could be considered. In the case of 1 Bligh Street in Sydney, only the central circulation core is naturally ventilated. © DEXUS Property Group

**Buildings in hot, dry climates should be oriented with the main façade openings positioned toward the north and south to reduce solar gain during periods of lower-angle sun in the mornings and afternoons.**

to be naturally ventilated for the whole year or simply during periods of the year when the climate is favorable. The following sub-sections outline some of the specific considerations relating to each distinct climate type.

### Tropical Climate

Tropical climates are characterized by relatively high air temperatures (although not as high as "hot dry" climates), high humidity levels, high precipitation, and high solar radiation. The high humidity – 80 percent or higher for at least five months of the year – means that this climate can be one of the most challenging to adopt natural ventilation strategies in, especially for office buildings.

Two important strategies can be utilized when designing in this climate: protection from solar radiation and

ventilating with a high air change rate to remove unwanted humidity. A high air change rate can also broaden the human comfort zone, as air movement over the skin creates a cooling sense (psychological cooling).

Tall buildings can also be a positive typology in the urban scale for hot, humid climates in that they often provide a positive effect on the surrounding microclimate, through shade and increased wind movement. Overshadowing and wind turbulence (inherent to tall buildings, especially at the base) improve comfort conditions in the surrounding urban context.

A consideration when designing in this climate is the small temperature difference (6–8 °C) between optimal internal temperatures for human comfort and external temperatures as the later are rarely greater than 32 °C. For this reason, solar heat gain on the building's façade needs to be carefully controlled. Although shading devices are important to be used, they should be carefully placed so as not to block wind/ventilation access. Even though day and night temperatures typically do not vary greatly (rarely greater than 8 °C), night-time ventilation can still be strategically applied to remove some daytime heat gain. Care should be taken, however, as humidity levels can remain high at night and can thus impact the effectiveness of night-time ventilation.

In this climate, the natural ventilation strategy will likely have a major impact on the form of the tower. Narrow floor plates are useful to ensure adequate cross-ventilation across the space. High floor-to-ceiling heights can be utilized to keep warmer air stratified upwards, away from the occupants, and also increase natural ventilation through stack effect. Menara UMNO in Malaysia (see pages 50–57) is a good example of how increasing the air change rate can help create a better internal environment through the process of psychological cooling. By sculpting the outside form and skin of the building, vertical "Wing Walls" were formed, which capture a wider angle of incident wind and accelerate it through a narrowing gap into the space, across the space, and out the other side. It should be noted, however, that higher air change rates can cause problems of discomfort and troublesome movement of papers, especially in office environments. It is not fully clear how effective the Wing Wall strategy at Menara UMNO is, nor what percentage of the year it is utilized. This lack of understanding of actual performance is exacerbated by the building being a multi-tenant building, with differing tenants utilizing differing strategies for ventilation (mechanical or natural) on each floor.

If the feasibility of having a fully naturally ventilated building in a tropical climate does not seem possible, another solution is to separate the building into natural and mechanical ventilation zones, with separate consideration of spaces that require a high degree of environmental consistency (e.g., office work spaces), and those that could tolerate more variation (e.g., circulation and social/gathering spaces). Although in a temperate climate, 1 Bligh Street in Sydney (see pages 132–137), is a good example of a building which provides natural ventilation in the main lobby, central circulation areas on each floor, and break-out meeting spaces, through the use of stack effect in a large vertical atrium supplied with fresh air at the ground level (see Figure 3.3) and extracted at height. This helps reduce the overall building cooling loads in a hot, humid climate.

### Hot, Dry Climate
Hot, dry climates are characterized by high air temperatures that surpass 37 °C for much of the year, low humidity levels, almost no precipitation, and ample day/night temperature fluctuations. Mitigating solar heat gain is an essential strategy in this climate. Buildings should ideally be oriented along an east–west axis with the main façade openings positioned toward the north and south. This orientation reduces solar gain during the more problematic lower-sun angle periods in the mornings and afternoons, especially in summer.

Natural ventilation during the day can be difficult in this climate due to the high external air temperatures and also the quantity of particulates in the air (especially sand). Some locales may offer an optimal environment for a seasonal switchover mixed-mode strategy. During the hotter months, night-time ventilation can be an effective strategy; utilizing the often significant drop from day-to-night-time air temperatures to flush out internal heat gains built up during the day. In this way, an office building can be cooled each evening before the work cycle begins the next day. A heavy thermal mass can delay thermal exchanges between the exterior and interior environments and help limit internal heat gains.

Evaporative cooling strategies coupled with a continuous airflow could also be used to increase the moisture content in the typically dry air, lower the temperature, and improve internal comfort. Evaporative strategies include localized evaporative cooling within the façade system or the use of a large centralized water source to help condition a specific area such as a water feature/fountain in the lobby of a building.

### Temperate Climate
Most of the case studies covered in this guide are located in temperate climates and are characterized by a clear distinction in ventilation strategy between summer and winter (or cooling and heating seasons). Three case study locations, however – Guadalajara, San Francisco, and Sydney – have less temperature variation between summer and winter and thus a more consistent climate year-round.

A temperate climate typically requires the architectural design to have the greatest adaptability throughout the year, employing devices such as high thermal insulation and passive heating during the cold season, but shading and higher ventilation rates during the hot season. This requirement for a level of adaptability of the systems in the building requires a careful study of the daily, seasonal and yearly variations of the local climate. Such adaptable strategies will typically include controlled ventilation that varies according to external conditions, and shading that reduces solar radiation in the summer but can allow it in winter.

Considerations such as wall-to-floor ratios, window-to-floor ratios, and building orientation will help determine the success of a natural ventilation strategy for a building in a temperate climate. In addition, the use of double-skin façades has become a more common strategy for naturally ventilating tall office buildings, especially in Europe (see Section 3.7 Façade Treatment and Double-Skin).

### Cold Climate
Cold climates are characterized by low average air temperatures (lower than −3 °C during winter) and low solar radiation as most are located above 40° north latitude. However, some cold climates can also have hot, humid summers, such as Winnipeg, Canada.

The most important consideration when designing in such extreme climate conditions is the conservation of heat. This can be addressed through both the form of the building and the building envelope. Compact shapes provide more concentrated floor-to-envelope area, which reduces the façade heat loss/gain (but may reduce the potential for natural lighting). Curvilinear shapes also have a better aerodynamic performance which may assist in the natural ventilation strategies in the building. The building should look to benefit from passive solar heating, especially in winter, by orienting the glazed areas of the façade toward the more intense solar radiation. Special attention, however, is needed when specifying the glazing components – considering the low thermal resistance of glass. The use of double and triple glass panels with gas-filled air cavities has become common practice in cold climates.

The benefits of passive solar heating can pose an issue with regard to solar glare at the working surface height, particularly in office buildings. The use of adjustable shading devices in the work spaces, coupled with occupant control, can address this concern.

Although rarely used in cold climates due to potential heat loss, natural ventilation can be used during the warmer shoulder months. With appropriate strategies and building orientation, as exemplified in Manitoba Hydro Place in Winnipeg, Canada (see pages 112–121), natural ventilation can be further extended beyond the transitional seasons. In this building, large expanses of glazing provide solar heat gain to the sky gardens at the south side of the building, which tempers the air supplied to the office spaces located along the west and northeast sides of the building. The air is then extracted via a second system of stacked vertical sky gardens on the north side, which link directly to a full-height solar chimney utilizing stack effect. A water feature within the southern lobby also helps humidify the incoming dry air in winter (and is chilled to help dehumidify it in summer), since this is another problem with natural ventilation in cold climates.

## 3.3 Site Context, Building Orientation, and the Relative Driving Forces for Natural Ventilation

The careful orientation of a tall building in relation to the prevailing wind and sun significantly improves the prospects of natural ventilation. Manitoba Hydro Place in Winnipeg, Canada (see pages 112–121) is a prime example of how the building form and massing responds to both solar orientation and the prevailing wind direction (see Figure 3.4). The building opens up toward the south, revealing a series of sky gardens which capture both Winnipeg's abundant sunlight and its consistent southerly winds. This orientation improves the effectiveness of natural ventilation and preheats the incoming fresh air, ideal for natural ventilation in cold climates.

Consideration should also be given to the primary driving forces which induce airflow in and out of the building. Although the predominant forces for naturally ventilating a tall building are likely to be bouyancy-induced, if wind (cross-ventilation) is the primary driving force, it is important to orient the main windward openings in the direction of the prevailing wind, as in the case of the San Francisco Federal Building (see pages 104–111). When there is little wind or the building openings are not able to be oriented in the direction of the prevailing wind, aerodynamic elements such as the use of wind Wing Walls in Menara UMNO, Penang, Malaysia (see pages 50–57), can be utilized to capture the wind through a wider incident angle and induce more effective natural ventilation.

▲ Figure 3.4: Manitoba Hydro Place is a prime example of how a tall building's form and massing can respond to both solar orientation and prevailing wind. © KPMB Architects/Gerry Kopelow

▲ Figure 3.5: View into the west façade which doubles as a thermal flue in the GSW Headquarters building in Berlin. © Annette Kisling

The prospects of a purely naturally ventilated building are significantly enhanced when the two driving forces, wind and buoyancy, act in unison.

Furthermore, a circular tower allows for better wind-induced ventilation from various directions.

Conversely, when both wind and buoyancy driving forces are employed simultaneously, there is greater flexibility with building orientation with regard to the prevailing wind direction. For example, the main windward openings in GSW Headquarters, Berlin (see pages 64–73) are located on the east side despite the wind predominately blowing from the west. GSW building's orientation is based on the sun path and its relation to the thermal flue (west-facing space) which utilizes stack effect induced by the afternoon sun. In this case, stack effect and solar heat gain create bouyancy-induced cross-ventilation and maintain sufficient airflow rates and comfort levels, even on hot days in the absence of wind.

Strong winds from a predominant direction can thus be exploited to allow for single-sided and/or double-sided cross ventilation. KfW Westarkade in Frankfurt (see pages 122–131) benefits from an average wind speed of four meters per second, which comes from the southwesterly direction for

approximately 50 percent of the year. The building's directional, aerodynamic form, and orientation allows fresh air to enter the exterior ventilation flaps of the double-skin façade. When using wind-driven natural ventilation, especially in tall buildings, the risk of excessively high winds at height needs to be evaluated and mitigated.

Based on the case studies in this report, the prospects of a purely naturally ventilated building are significantly enhanced when the two driving forces, wind and buoyancy, act in unison. For tall buildings to rely exclusively on natural ventilation, an inward flow of air must be maintained even when winds are weak, there is no wind, or the wind is not blowing from the desired direction. With little or no wind, buoyancy-induced ventilation can provide an alternative/additional driving force to ensure the effectiveness of natural ventilation. As explained in the case studies, stack effect can be exploited through the use of an interior atrium or sky garden (see Section 3.5 Sky Gardens and Vertical Segmentation of Atria), through a thermal flue or solar chimney, or through vertical circulation cores.

GSW Headquarters, Berlin (see pages 64–73) and Manitoba Hydro Place, Winnipeg (see pages 112–121) demonstrate the exploitation of stack effect through the use of a thermal flue (see Figure 3.5) and solar chimney, respectively. Solar heat augments temperature and pressure differentials between the bottom (inlets) and top (outlets) of the stack, thus increasing the rate of upward air movement and inducing the fresh air entry through the inlets. This can induce cross-ventilation through spaces venting onto the atrium/thermal flue, allow for greater control over natural air movement, and maintain sufficient ventilation rates and comfort levels even on hot days in the absence of wind.

Additionally, thermal flues and solar chimneys present an opportunity to recover heat from exhaust air as demonstrated in Manitoba Hydro Place. It should be noted that using heat-recovery equipment can impede the flow and reduce flow volumes. One such solution is to have two exhaust routes for the flue: one in winter that passes through heat-recovery equipment when less natural ventilation air is necessary, and a second for summer

**In addition to bringing daylight deeper into the building plan and assisting natural ventilation, sky gardens also allow visual, social, and physical connectivity as a destination and transition space within a tall building.**

▲ Figure 3.6: Typical office space in the Post Tower in Bonn, Germany, which enjoys high floor-to-ceiling heights due to its decentralized mechanical system. © Murphy/Jahn Architects

without heat-recovery for when it is not necessary but higher airflow rates are essential.

Liberty Tower of Meiji University, Tokyo (see pages 42–49) exploits vertical circulation elements (specifically an escalator void) as an air extraction chimney utilizing stack effect in the void. In this scenario, the heat source is internal gains, rather than solar gains, since the escalator void is located deeper within the floor plan. The circulation element has surpassed its initial function as only a utilitarian element and became an integral component of the natural ventilation strategy. When implementing such a strategy, care should be given to both smoke control and providing safe egress to occupants during a fire (see Section 3.10 Fire Engineering / Smoke Control).

### 3.4 Planning and Spatial Configuration

The planning and spatial configuration of a tall building largely determines the possibility and effectiveness of natural ventilation. Cellular offices can benefit

from stack effect in an adjoining atrium while an open-plan office may be more conducive to cross-ventilation. Additionally, the floor-to-ceiling height can have a major impact on the natural ventilation effectiveness. The British Council of Offices Guide recommends a minimum floor-to-ceiling height of three meters (2.7 meters is closer to the standard) to enable more airflow through the interior space (Gonçalves 2010). With the removal of a centralized mechanical system and the utilization of a decentralized mixed-mode system, Post Tower, Bonn, Germany (see pages 74–83) operates without the necessity of multiple mechanical floors. In doing so, floor-to-ceiling height increases while the floor-to-floor height decreases compared to a traditional system with dropped ceilings and ducts (see Figure 3.6).

The case studies outlined in this technical guide have tended to challenge other traditional planning strategies, such as conventional lease depths and the preferred arrangement of cellular and open-plan office spaces. Prior to the advent of air-conditioning, high-rise buildings in the late 1890s

and early 1900s utilized limited plan depths combined with open (central) courts to exploit natural ventilation. Iconic high-rises such as the Chrysler and Empire State Buildings illustrate the importance of wall-to-core depth (limited to 8–9 meters during that time) to allow for sufficient daylight and natural ventilation. Current US practice suggests limiting floor plan depths to 7–8 meters from a window in order to be considered a naturally ventilated space. While projects such as GSW Headquarters, Berlin (see pages 64–73) boast a remarkably narrow plan depth range of 7.2–11 meters, there is still a conflict between the desired deep lease depth of commercial developers and the necessity of shallow floor plates to provide sufficient daylight and natural ventilation.

A few case studies, such as Commerzbank, Frankfurt (see pages 32–41) with a plan depth of 16.5 meters and Deutsche Mess AG, Hannover, Germany (see pages 58–63) with a plan depth of 24 meters, are able to provide natural ventilation with more desirable deep lease depths. Deutsche Messe AG located the circulation cores to opposing corners of the building to achieve this, while Commerzbank utilized a central atrium and sky gardens as well as moving the circulation cores to the edges of the building. In both scenarios, building occupants are not further than 8–12 meters from sufficient daylight and natural ventilation and the relocation of the vertical circulation elements and service facilities allow for unobstructed flow of fresh air across the office spaces.

Several case studies in the guide also demonstrate that the implementation of natural ventilation can require a radical change in the interior layout from that of a conventional tall office building (which typically sees cellular offices along the perimeter and open-plan office spaces at the center). The open-plan office spaces of both Manitoba Hydro Place, Winnipeg (see pages 112–121) and the San Francisco Federal Building (see pages 104–111) are located at the periphery of the building, with the conference rooms, private meeting spaces and cellular offices located toward the center of the floor plate . Both scenarios utilize low partition walls in the open-plan offices and gaps between enclosed rooms and the ceilings to permit airflow from one side of the office space to the other with minimal obstructions. Conversely, Post Tower has located its cellular office on the perimeter of the building in a traditional fashion, but utilizes "loose fittings" at the glass partitions to allow air to flow through the offices to the corridor beyond.

## 3.5 Sky Gardens and Vertical Segmentation of Atria

The use of sky gardens in the design of naturally ventilated tall buildings has now become quite common (see Figure 3.7). Approximately half of the case studies profiled in this guide

employ some form of a communal sky garden space. From a natural ventilation standpoint, the sky gardens can be used for air intake, air extraction, a combination of the two, or to induce ventilation in inward-facing offices, as is the case at Commerzbank, Frankfurt (see pages 32–41). In this building, a central atrium coupled with the radial arrangement of offices and sky gardens has provided great flexibility in ventilating the building, irrespective of prevailing winds.

In most scenarios, sky gardens are used as extraction chimneys which exhaust air from the building through stack effect. In Manitoba Hydro Place (see pages 112–121), however, south-facing winter gardens provide air intake and pre-heat incoming cold air during the heating season. This building also has north-facing atria that exhaust stale

air from the office spaces into the solar chimney. In all cases, sky gardens function as buffer zones which mediate the temperatures between exterior and interior. When they are located at the exterior of the building and used as air intakes, although the scale of the spaces and environmental configurations are obviously much different, they conceptually offer some of the same benefits presented by double-skin glass façades, such as thermal insulation and protection against undesirable weather conditions, noise, and high wind speeds.

In addition to bringing daylight deeper into the building plan and assisting natural ventilation, sky gardens also allow visual, social, and physical connectivity as a destination and transition space within a tall building. One can argue that, with increasing

▲ Figure 3.7: The Commerzbank building in Frankfurt, Germany houses sky gardens every four floors as part of the building's natural ventilation strategy. © Nigel Young / Foster + Partners

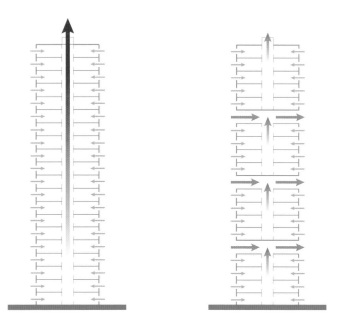

▲ Figure 3.8: An unsegmented tall building (left) can create extreme stack flows. This can be alleviated by introducing segmentation in the building (right).

▲ Figure 3.9: The cylindrical form of 30 St. Mary Axe facilitates the potential for natural ventilation strategies within the building. © Steven Henry/CTBUH

inner-city densification, population increases, and the consequent loss of open spaces (the public realm) at the ground level, tall buildings will increasingly require communal sky gardens as part of a more "public" program in the sky. As a destination space, sky gardens provide a hospitable environment for social interaction and recreation. As a transition space, sky gardens function like sky lobbies, providing ease of movement and allowing occupants to orient themselves within the building and urban context (Pomeroy 2008).

As stack effect is complex and can be seen as a problem in more extreme climates, it warrants careful analysis. Vertical segmentation (in an atrium or shaft of a high-rise building) prevents the development of extreme stack flows which may cause excessive drafts and occupant discomfort (see Figure 3.8). When the atrium or shaft is the primary source for fresh air, segmentation can reduce the variation of airflow rates between different floors.

If sky gardens are a key component of the natural ventilation strategy, the segmentation can occur by locating a series of sky gardens on top of each other. Post Tower in Bonn, Germany (see pages 74–83) utilizes this strategy, with four stacked sky gardens to exhaust the building. The central atrium of Commerzbank is also a key component in its natural ventilation strategy. The atrium, which runs the full height of the building, is segmented into 12-story "villages" with steel and glass diaphragms. Segmentation is not possible with the use of a thermal flue or solar chimney, however, where the full interrupted space of the building is needed. In this scenario, automatically controlled dampers can be used as in GSW Headquarters, Berlin (see pages 64–73), to control the pressure differentials within the flue/chimney.

Besides reducing the risk of creating large pressure differences and excessive drafts, there are other benefits to segmenting a vertical void into multiple sky gardens or atria. In the case of a fire, smoke can be mitigated to one specific segment of a tower. While 30 St. Mary Axe, London (see pages 84–91) is a multi-tenant building, segmentation also allows tenants to be visually and acoustically separated from other tenants in the building. Segmentation can also provide multiple social spaces in closer proximity to the occupants as in Commerzbank, which has one sky garden for every four floors.

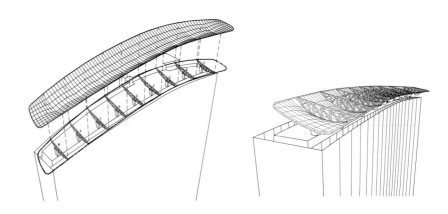

▲ Figure 3.10: A "Wing Roof" located directly over the west-facing thermal flue of the GSW Headquarters, Berlin, uses the Venturi effect to generate an additional uplift force in the thermal flue that help exhausts the building. © Sauerbruch Hutton

## 3.6 Aerodynamic Elements and Forms

The use of aerodynamic elements and forms can enhance the possibility of designing a purely naturally ventilated tall building. The external form of a building can optimize aerodynamics and enhance the flow of wind around it. The cylindrical form of 30 St. Mary Axe, London (see pages 84–91) encourages wind to accelerate as it goes around the building, creating pressure differentials between the windward and leeward sides (see Figure 3.9).

Additionally, aerodynamic elements can be applied to the tower to facilitate the natural ventilation strategy. GSW Headquarters, Berlin (see pages 64–73) uses a "Wing Roof" located directly over the west-facing thermal flue (see Figure 3.10). This aerodynamic element is shaped in profile like an upside-down airplane wing and uses the Venturi effect to generate an additional uplift force in the thermal flue that help exhausts the building. When the wind is not blowing in the prevailing direction, a series of fins suspended from the Wing Roof causes the wind to eddy, which prevents the risk of down currents over the outlet of the flue.

Within the relatively calm tropical winds of Malaysia, Menara UMNO (see pages 50–57) uses wind Wing Walls to direct winds into the building from a wider range of directions than the prevailing wind. Wing Walls also capture and create greater positive pressure on the windward side of the building, so the natural ventilation strategy does not need to rely solely on the naturally created negative pressures on the leeward side of the building to draw ventilation through. In theory, this induces a high air change rate necessary to achieve thermal comfort conditions in a tropical climate (see Section 3.2 Local Climate). However, one could raise the hypothesis that if the Wing Wall device was used in a location with higher average wind speeds, the pressure differentials across the building envelope could be too great, causing problems such as difficulties with opening windows and doors and controlling the high airflow rates.

While some buildings create large aerodynamic elements, other buildings employ smaller elements to induce natural ventilation. RWE Headquarters Tower in Essen, Germany uses a "Fish

**Wing Walls can capture and create greater positive pressure on the windward side of a building, so the natural ventilation strategy does not need to rely solely on the naturally created negative pressures on the leeward side of the building to draw ventilation through.**

"Mouth" device to adjust the speed of air intake through the double-skin façade (see pages 24–31). A similar, yet reversed, device is used to exhaust warm air from the façade cavity. Commerzbank, Frankfurt (see pages 32–41) similarly uses small aerofoil sections at the top/bottom of the double-skin façade openings to improve airflow through the ventilated cavity and avoid the short-circuiting of air.

## 3.7 Façade Treatment and Double-Skin

After the decision to naturally ventilate tall buildings is made, the façades need to be carefully designed to respond to the specifics of air speed, urban noise, noise generated by air movement, and solar protection, among other things. As this guide has shown, there has been a tendency in recent years – especially in Europe – toward employing double-skin façades to assist with natural ventilation. There are numerous benefits of such twin-layer envelopes, however, selecting the appropriate glazing is crucial to getting the necessary performance.

In a double-skin façade, the outer glazing layer provides additional protection against weather conditions and external noise (both urban level and wind induced) and allows greater control of incident wind speeds, therefore enhancing the prospect for natural ventilation in high-rise buildings. Additionally, a double-skin façade acts as a thermal buffer, mediating temperatures between the interior and exterior.

During the heating season, solar gain in the cavity of the double-skin can serve to preheat fresh air before it enters the building. Conversely, during the cooling season, the cavity is ventilated in order to carry away solar heat gain. A double-skin façade also allows for the integration (and protection) of shading devices within the cavity, which provides an added benefit of reflecting solar heat gain before it enters the building. The ventilated cavity can thus exhaust the solar heat in the cavity before it has the opportunity to penetrate the inner façade layer and inhabited spaces beyond.

Careful consideration should be given to the design of a double-skin façade. If the cavity is not sufficiently ventilated and/or sunshading devices are not properly positioned, the façade could experience overheating in the summer. Furthermore, double-skin façade design requires specific considerations for

| | Commerzbank, Frankfurt, 1997 | RWE Headquarters Tower, Essen, 1996 | 1 Bligh Street, Sydney, 2011 | KfW Westarkade, Frankfurt, 2010 | GSW Headquarters Tower, Berlin, 1999 | 30 St. Mary Axe, London, 2004 | Manitoba Hydro Place, Winnipeg, 2008 | Deutsche Messe AG Building, Hannover, 1999 | Post Tower, Bonn, 2002 |
|---|---|---|---|---|---|---|---|---|---|
| **Cavity Depth** | 200 mm | 500 mm | 600 mm | 700 mm | 1,000 mm (west façade) 200 mm (east façade) | 1,000–1,400 mm | 1,300 mm | 1,400 mm | 1,700 mm (south façade) 1,200 mm (north façade) |
| **Horizontal Continuity** | 1.5 meters | 2 meters | Fully Continuous (around entire perimeter) | Fully Continuous (around entire perimeter) | Fully Continuous (along entire length of façade) | Varies (between diagonal structural frame members) | Fully Continuous (along entire length of façade) | Fully Continuous (around entire perimeter) | Fully Continuous (along entire length of façade) |
| **Vertical Continuity** | 2.4 meters (between floor spandrel panels) | 3.5 meters (floor-to-floor) | 3.85 meters (floor-to-floor) | 3.7 meters (floor-to-floor) | Fully Continuous Approximately 67 meters (full height of façade) | 4.15 meters (floor-to-floor) | 4 meters (floor-to-floor) | 3.2 meters (floor-to-floor) | Approximately 32 meters (sky garden height) |

▲ Figure 3.11: Table comparing the double-skin façades of the case study buildings in this Technical Guide.

the position and sizing of openings to ensure proper pressure distribution within the natural ventilation system. High-rise buildings experience significantly different wind speeds and pressures differentials at various heights and locations across the façade, and the double-skin openings need to somehow account for these. The Fish Mouth device in the double-skin façade of RWE Headquarters Tower in Essen, Germany (see pages 24–31) changes in size above floor 16 to moderate wind pressures across the envelope openings. At KfW Westarkade, Frankfurt (see pages 122–131), the sophisticated BMS system controls the double-skin façade to act as a complete "Pressure Ring" altering the degree of openness of exterior ventilation flaps to control and maintain the pressure differentials across the façade.

Nine of the fourteen case studies in this guide, six of which are located in Germany, utilize a double-skin façade. It should be noted, however, that no two buildings employ the same solution for the double-skin/natural ventilation strategy (see Figure 3.11). Even the depth of the cavity differs significantly across these double-skin case studies, from 200 to 1,700 mm, with some buildings having varying cavity depths across the differing faces of the same building. Although there seems to be good consensus that double-skin façades have a potentially positive role to play in the natural ventilation of tall office buildings, especially in more temperate climates, there is not yet consensus on what detailed form the double-skin should take. All the offset factors also need to be taken into account – such as increased material and construction costs or loss of floor space. Perhaps this is why the strategy has yet to gain any traction in the more commercially orientated US building industry.

▲ Figure 3.12: Single-skin façade of Highlight Towers in Munich is ventilated through operable panels behind perforated, fixed steel panels. © Murphy/Jahn Architects

While numerous buildings in this technical guide use double-skin façades, there are other examples included which challenge its necessity. The Highlight Towers, Munich (see pages 92–97) have a single-skin façade and are naturally ventilated through pivoting panels integrated in the façade. Fixed, triple-glazed windows feature high-performance and heat-reflecting properties while a narrow, operable glass panel enables direct natural ventilation. A perforated, stainless steel panel (with soundproofing properties) is mounted behind the pivoting panel to provide protection against the sun, wind, and rain (see Figure 3.12).

In hot climates, a predominantly enclosed double-skin façade could possibly be detrimental to natural ventilation due to the risk of the façade cavity overheating. Torre Cube in Guadalajara (see pages 98–103) takes this into account by utilizing a more open, exterior wooden lattice,

**Careful consideration should be given to the design of a double-skin façade. If the cavity is not sufficiently ventilated and/or sunshading devices are not properly positioned, the façade could experience overheating.**

▲ Figure 3.13: Torre Cube, Guadalajara uses a simple, wooden lattice brise-soleil to give shade and control solar gain. © Estudio Carme Pinós/Lourdes Grobet

**The performance of natural ventilation can benefit by exposing thermal mass, utilizing night-time cooling, minimizing solar heat gains, optimizing the façade, and providing radiant cooling/ heating.**

brise-soleil screen to protect again high wind speeds, wind-driven rain, and solar heat gain. Since this lattice screen is open on all sides, air can pass through the interstitial space, mitigating the risk of excessive heat buildup and allowing convective cooling on the glass façade.

### 3.8 Related Sustainable Strategies

Since the primary purpose of natural ventilation is to provide acceptable air quality and human comfort, the use of other sustainable strategies which complement natural ventilation can help reduce the cooling load and enhance comfort conditions. As demonstrated by several case studies in this guide, the performance of natural ventilation can benefit by exposing thermal mass, utilizing night-time cooling, minimizing solar heat gains, optimizing the façade, and providing radiant cooling/heating. Load reduction

strategies, including internal lighting and solar heat loads, is the number one priority for assisting natural ventilation. By reducing these gains, natural ventilation as an effective cooling strategy can be utilized for a greater period of time.

Night-time ventilation can be an efficient passive strategy in office towers because it takes advantage of lower external air temperatures during the night to cool down the building and purge the internal heat loads gained throughout the day, which are stored in the fabric of the building. This prospect can be enhanced by exposing more thermal mass for the storage of heat, such as through exposed concrete slabs. The average 11 °C day/night temperature difference in Bonn allows Post Tower to effectively utilize night-time ventilation in summer, providing a cooled office space for the following workday (see pages 74–83). It is important to note that night-time ventilation is only effective in certain climates that experience significant day/night temperature differences. Additionally, a high relative humidity in the climate could negatively affect this strategy.

Solar shading devices that maintain good daylight penetration quality are highly important since high solar heat gain can overwhelm a natural ventilation system. Contrarily, in the warming season, a building can benefit from passive solar heat gain to preheat incoming air. An adaptable strategy should be put in place to limit solar heat gain when necessary, allow for solar gain when beneficial, and always reduce glare from occupants' work spaces. Further, the optimization of daylight can minimize the dependence on artificial lighting, which contributes significantly to energy consumption in the building (typically 30 percent or more of the total building energy consumption) and also adds significantly to internal

▲ Figure 3.14: View looking up at the RWE Tower in Essen, Germany with its ultra-clear glass façade. © ingenhoven architects

▲ Figure 3.15: CFD modelling was used to predict airflow patterns around the Commerzbank building in Frankfurt, Germany, which was one of the first major commercial buildings to employ CFD. © Foster + Partners

heat gains. Torre Cube, Guadalajara (see pages 98–103) uses a simple, wooden lattice brise-soleil (see Figure 3.13) that can be manually operated by the individual occupants to give shade and control solar gain. RWE Headquarters Tower in Essen, Germany (see pages 24–31) boasts the use of clear glass by incorporating perforated blinds within the double-skin cavity, and an alternate interior anti-glare screen (see Figure 3.14).

Several case studies in this guide demonstrate how they have supplemented their cooling/heating needs with radiant slabs. Radiant slabs involve the circulation of chilled or warmed water in pipes or panels that are embedded in the ceiling or floor to exploit the surface's thermal storage capacity. A major benefit is that radiant slabs consume less energy than air-based forms of heating/cooling as it is more efficient to warm/cool a liquid than it is air. The employment of radiant cooling in conjunction with natural ventilation can significantly reduce, or even

eliminate, the need for air-conditioning. Internal loads and the relative humidity levels of the climate are important factors to be considered, as there is a risk of condensation being produced on chilled surfaces.

## 3.9 Predictive Performance and Modeling

Since there are far more variables in a natural ventilation strategy, there is less certainty of delivery than with a traditional mechanical ventilation system. It is even more important to adequately model and predict the ventilation performance prior to final design and construction. There are several techniques that can be used to aid the design of naturally ventilated buildings, with wind tunnel testing, computational fluid dynamics (CFD), and salt bath modeling being the most common. Salt bath modeling is an experimental technique where scale models (between 1:20 and 1:100) are inverted and placed in a large tank of

fresh water. A dyed brine solution is injected into the model to represent sources of heat and allow buoyancy-driven airflow to be visualized. This strategy does have disadvantages, such as the time and expense to create accurate models and the lack of simulated heat exchange with the structure; thus wind tunnel testing and CFD modeling are more common and relied on.

Ideally a combination of both wind tunnel testing and CFD analysis would be used to assess wind impacts, since CFD simulations alone may not accurately evaluate wind-driven ventilation. At the design stage, in-depth CFD simulations (see Figure 3.15) and wind tunnel testing can be carried out to predict airflow patterns, indoor air velocities, indoor temperature distribution, and ventilation rates associated with different combinations of window openings/configurations and/or various outdoor wind speeds and weather conditions. The result of these tests and simulations should then be used to inform the configuration and sizing of façade apertures such as window openings. In Menara UMNO, Penang, Malaysia (see pages 50–57), the Wing Wall device was introduced as a direct result of the CFD simulations, which indicated that high airflow rates could be achievable if the Wing Walls were provided. CFD simulations also showed indoor airflow patterns which could be used to inform the interior layout and spatial configuration of the offices.

At a more detailed level, CFD simulations led to modifications in the user-controlled windows in the San Francisco Federal Building (see pages 104–111). By introducing an inflow deflector that disrupts the short-circuiting and recirculation of air, these studies demonstrated how small improvements in the detailed design of building elements can have a significant impact on flow distribution and the overall efficiency of the ventilation system. CFD simulations also test the effectiveness of natural ventilation in extreme temperatures and in "worst-case" locations in the building to ensure that comfort is maintained in all zones during occupied hours.

## 3.10  Fire Engineering and Smoke Control

The principle fire safety strategy for tall buildings is that vertical fire and smoke spread should be resisted, and contained to the floor of fire origin. It is common for the natural ventilation aims of a building design to conflict with these fire safety objectives. While stack effect can be a great strategy to enhance natural ventilation, it can also provide a powerful distribution force for a fire. Rising heat from a fire in an enclosed stack will draw in fresh air at increased speeds and thereby contribute to the growth and spread of the fire. Natural cross-ventilation is not subject to the same magnitude of risks

as the smoke/fire can be more strictly contained in one area, rather than spread vertically across several floors as with stack effect ventilated spaces.

A naturally ventilated building can challenge the operational aspects of mechanical smoke control systems and pressurized stair systems that generally rely on tightly closed façades in order to develop the pressure differentials and appropriate airflow patterns necessary to manage or resist the movement of smoke and fire within a tall building. While different jurisdictions have different requirements with regard to smoke management in tall buildings, some countries actually require operable windows as part of the fire engineering strategy. In some cases the smoke management strategy may dictate the natural ventilation solution. Despite the potential conflict between natural ventilation and fire safety objectives, the challenges can be adequately addressed through thoughtful design.

Ventilation solutions which use double-skin façades create a fire spread route which does not exist on buildings with less complex, fixed façades and consideration must be given to the migration of smoke through the cavity and back through openings into a non-fire affected floor. The risk of fire spread through double-skin façades introduces concerns regarding if the flame should break through the inner façade and then be confined within a long tall shaft-like space. The dynamics of the confined flame and radiant heat exposure are potentially more severe than a flame freely flowing to the open atmosphere.

Other types of double-skin façades may reduce the risk of fire spread, particularly those using a partitioning scheme within the cavity. The double-skin façade of KfW Westarkade, Frankfurt (see pages 122–131) is segmented on a

floor-by-floor basis and would thereby contain a fire within its floor of origin (see Figure 3.16). Additional segmentation of the façade cavity occurs in the horizontal plane in relation to the positioning of internal fire partitions.

The materials used to construct a curtain wall can be an important consideration in the relative risk to fire spread vertically up or down the building face. Façades are often constructed primarily of metal and glass materials, but it is also common to find architectural cladding systems comprised, in part, of combustible materials. Examples include aluminum-faced sandwich panels with a polyethylene core, or combustible exterior insulation and finish systems (EFIS), among numerous others. The use of these combustible materials, although qualified by fire test standards, may require additional considerations if used in the context of a ventilated façade.

Naturally ventilated buildings tend to be highly individual and there are few common solutions to the conflicting demands of natural ventilation and fire separation in tall buildings. Typical solutions to achieve fire safety in a naturally ventilated building may include (but are not limited to):

(i) automatic dampers to restore compartmentalization in a fire event,
(ii) window sprinkler systems,
(iii) additional ventilation to clear or dilute smoke,
(iv) mechanical smoke extraction system,
(v) pressurization systems, and
(vi) vertical or horizontally mounted sliding fire doors/screens.

The use of mechanical extraction and pressurization systems may be exceedingly more difficult to design given the variable operations of a ventilated façade in wide-ranging climates.

It is common for the natural ventilation aims of the building design to conflict with the fire safety objectives. While stack effect can be a great strategy to enhance natural ventilation, it can also provide a powerful distribution force for a fire.

▲ Figure 3.16: Overall view of KfW Westarkade, Frankfurt, where the floor-by-floor segmentation of the double-skin façade would thereby contain a fire within its floor of origin. © Jan Bitter

Sprinklered high-rise buildings have a very successful record of life safety and property protection performance. For this reason, US building codes do not require fire resistance rated spandrels or flame deflectors at the building façade in fully sprinklered buildings. A competent fire and life safety engineer would minimize any systems so that the natural ventilation strategy and fire strategy are harmonized.

One possible aid to achieving harmonization is in the use of CFD models to better understand the fire and smoke spread risks. In addition to CFD models, it is often appropriate to carry out a risk assessment. The use of one or more risk assessment methodologies enables active fire safety solutions such as window sprinklers to be compared against passive solutions such as a fire resistant façade. This approach can

ensure that fire safety can be achieved and that the natural ventilation solution has not introduced a level of risk that would have not existed in a building without natural ventilation. Several factors to consider in a risk assessment of fire spread at the building façade include, but are not limited to, the following:

(i) automatic sprinkler systems' reliability,
(ii) fire department response capabilities,
(iii) building height,
(iv) building occupancy,
(v) building compartmentalization features,
(vi) building evacuation strategies,
(vii) fire hazards such as fuel loads, continuity of combustibles, compartment sizes, etc., and
(viii) security threat assessment scenarios.

### 3.11 Other Risks, Limitations, and Challenges

#### Technical Sophistication and Operability

Even though natural ventilation is generally considered a passive, low technological approach to condition a building, it can require highly technical systems to monitor and control the building. In a mixed-mode building, sensors are used to measure internal environmental conditions and external conditions of air temperature, humidity, wind velocity, and rain. The BMS will decide, based on this information, to utilize passive or active modes. With the use of a sophisticated control system, the BMS will need to be constantly fine-tuned to respond to variations in climate and occupant behavior. This is especially the case during the first two years of building operation. This

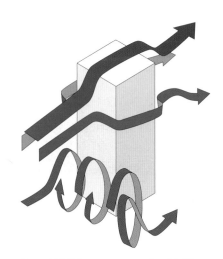

▲ Figure 3.17: Airflow pattern around a tall building will cause a vortex at its base.

▲ Figure 3.18: The vortex effect on a tall building can be mitigated depending on the form of the building, such as with a curvilinear building as is the case with as 30 St. Mary Axe in London. © Steven Henry/CTBUH

**If little, or no, attention is paid to envelope loads, lighting loads, or equipment loads in the office space, natural ventilation will not be effective in providing occupant thermal comfort.**

also implies the required presence of a full-time facilities manager and maintenance team on-site to control the operation of the BMS. Although the automatic control can provide the optimum comfort, giving the occupants manual control can improve user satisfaction. Care must be taken, however, that the actions of individuals do not alter the working of the system overall, especially for people located in other areas in the building. Occupants must be fully aware of how to operate the building and the systems need to be easy to use and understand. In most cases, an intelligent control panel will inform occupants which mode the building is operating in, yet allow the system to be overridden. Systems defaulting to natural ventilation but with a user override can provide a good system for improved occupant comfort and lower energy.

### Internal Load Reduction Strategies
While internal heat gains may be an added benefit in cold climates, most conditions will require the cooling strategy to compensate for additional

heat. If little, or no, attention is paid to envelope loads, lighting loads, or equipment loads in the office space, natural ventilation will not be effective in providing occupant thermal comfort.

### Airflow and Wind Patterns
The risks and challenges associated with natural ventilation in a high-rise building are considerably greater than in low-rise structures, since wind speed increases, and airflow patterns become less predictable at height. As air contacts the windward face of a building, approximately one-third of the air travels upward/over the building and the remainder flows downward, forming a vortex at the ground (see Figure 3.17). The vortex effect is amplified and more air passes around the sides and over the top of the structure as the building height increases. However, such challenges can be mitigated depending on the form of the building. A tall, slender rectilinear building or a curvilinear building (see Figure 3.18), such as 30 St. Mary Axe in London (see pages 84–91), provides a viable solution.

As tall buildings are typically located in dense urban areas (see Figure 3.19), unique micro-climatic conditions occur which can induce unusual wind patterns around the building. The complexity and influence of wind in the "urban street canyon" can pose a significant challenge when determining optimal building orientation, aerodynamic design, and the location of natural ventilation openings. Available data on prevailing winds are typically taken at airports, far away from city centers and the non-typical setting of the site. These unusual wind patterns can be influenced by several factors such as surrounding building heights, spacing between adjacent buildings, aerodynamic roughness, surrounding building geometries (e.g., height-to-width ratios, wall-to-plan area ratios, aspect ratio), building volume characteristics, urban heat island influences, etc. While it is difficult to model these complex wind patterns, several tools such as urban field measurements, wind tunnel studies, and CFD model simulations can help address the site-specific challenges.

▲ Figure 3.19: 1 Bligh Street in Sydney is an example of how tall buildings are typically located in dense urban areas which can create unique micro-climatic conditions which can induce unusual wind patterns around the building. © ingenhoven architects

### Maintenance
Just as with fully air-conditioned buildings, naturally ventilated buildings also require significant levels of maintenance. In particular, a higher degree of maintenance is typically required of the components that make up the operable façade. This typically includes calibration of sensors and actuators and routine replacement of moving parts.

As most high-rise office buildings are located in dense urban centers, high levels of air pollution may place restrictions on the use of natural ventilation. There is a perception that mechanical systems clean the air more extensively before it enters the building than with natural ventilation systems. However, after removing dust and particulate matter, there is often no further

cleaning to limit polluted air from entering the building. Other air-borne hazards include birds and small animals, sand (especially in hot, arid environments), insects, pollen, dust, odors, and fumes from vehicular traffic. Insects and other particulates require that a screen be provided over intakes to protect the openings. However, these screens may reduce ventilation rates and impose a heavy pressure loss on air movement if not designed correctly. They also need to be regularly cleaned.

With a natural ventilation system, the ingress of dust could have a negative impact on tenant perception and require higher levels of cleaning. Additionally, the inner faces of a double-skin façade will require regular cleaning to avoid dirt build up. Most, but not all, cases allow access to the double-skin cavity at every floor. Alternatively, the cavity width can often accommodate a person to enter and move around for cleaning purposes by catwalk or abseil.

### Acoustic Considerations
Connecting occupants with the external environment has some great benefits, but it can also expose occupants to unwanted noise from ambient airflow and urban street noise such as traffic or mechanical plant on adjacent buildings. This may even limit the amount of hours the windows (and possibly the whole natural ventilation system) can be operable. While double-skin façades can provide some acoustic isolation from unwanted noise, the solution of an acoustical, perforated panel over the openings in the single-skin façade of Highlight Towers, Munich is interesting (see pages 92–97). It is also important to understand that urban noise is likely to pose more of a problem on the lower floors of a building.

### Safety/Security
Openings at the lower levels of buildings can pose security risks, as seen at the San Francisco Federal Building (see pages 104–111). The US General Services Administration, which oversees the construction of US federal buildings,

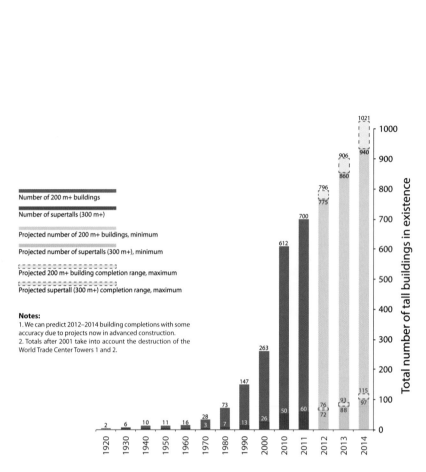

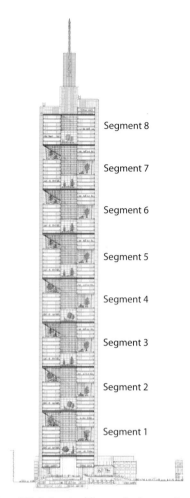

▲ Figure 3.20: Graph showing the significant increase in tall buildings over the last decade, with 700 buildings over 200 meters in existence as of the end of 2011. © CTBUH

▲ Figure 3.21: An imagined Commerzbank supertall building increasing the number of "villages."

stipulates that the first five floors of a federal office building cannot have operable windows. Thus, in the case of the San Francisco Federal Building, the first five floors are sealed and fully air-conditioned.

Consideration also needs to be given to the design of motorized openings. Liberty Tower in Meiji University, Tokyo (see pages 42–49) provides a protective screen over the motorized ventilation flaps to prohibit occupants from having their hands or fingers trapped in the closing vent. These exterior screens also prohibit small birds or other pests from entering the building. Consideration also needs to be given to the risk of objects falling through openings and harming people at ground level.

## 3.12 Looking to the Future: Naturally Ventilating the Supertall

Although tall buildings have been increasing in both height and number exponentially over the past decade or so (see Figure 3.20), there are currently no supertall buildings (greater than 300 meters in height) in existence that employ natural ventilation strategies. As the height of a tall building increases, new challenges arise in implementing a natural ventilation strategy. Air pressure and temperature differentials across the façade will increase, presenting a challenge in designing and sizing external envelope openings and devising a ventilation control strategy. Climate and wind speeds can be vastly different at increasing heights than at ground level.

Segmentation may offer a plausible solution to potentially allow for the

employment of natural ventilation for the supertall building typology. For example, each nine-story village in the Post Tower Bonn building, and 12-story village in the Commerzbank, Frankfurt behaves as an individual unit independent of the ventilation strategy of the segment above or below. As each segment addresses pressure differentials independently, the natural ventilation strategy could potentially be vertically extruded by adding additional segments to create the supertall building (see Figure 3.21).

As climate plays a major role in the decision to naturally ventilate a typical high-rise, so will it also play a major role in the feasibility for natural ventilation in a supertall building. The case studies in this technical guide have suggested that it is difficult to implement natural ventilation in a tropical climate.

However, a tropical climate might provide the ideal scenario for a Zoned strategy for a supertall building. With increasing heights, winds tend to be stronger and the air temperatures tend to be lower. A hot, humid climate with little to no wind speed at ground level may provide the optimum environment for natural ventilation at greater elevations.

## 3.13 Conclusion: Challenging Industry and Occupant Preconceptions

Adopting natural ventilation systems in not only tall office buildings, but many building types, faces a number of challenges beyond designing the most effective system for a particular site, program, building type and configuration. While there is the potential to greatly reduce energy costs, there is a perception that both construction costs and maintenance costs will be higher. When the potential "loss" of floor space through the use of double-skin façades and sky gardens or atria are factored in, it is not surprising that some developers consider the approach to be "too expensive" even when the long-term savings can be significant. However, in a mixed-mode building, European practice has suggested that 30 percent of yearly occupational hours utilizing natural ventilation will justify the mixed-mode strategy economically (Gonçalves 2010), but the perceived higher initial construction costs can be a difficult barrier to overcome.

What is increasingly becoming recognized in the whole sustainability equation is not just the savings through reduced energy/carbon bills, but the positive impact on employee productivity through the creation of a higher quality (and perhaps more natural) internal environment. This will perhaps have the biggest bearing on the greater adoption of natural ventilation for office buildings in the future, rather than the savings in energy consumption.

As an example of this, the Advanced Building Systems Integration Consortium (ABSIC) and the Center for Building Performance and Diagnostics (CBPD) has created the Building Investment Decision Support (BIDS), a cost-comparison analysis of high-performance buildings and the enhanced quality of the individual workplace (Loftness et al. 2004). This tool proposes that sustainability should not be considered solely on first construction costs. If approached more holistically, the sustainable environmental gains through a mixed-mode or natural ventilation system can decrease costs in three of ten major areas, including Individual and Organizational Productivity, Facilities Management, and Energy. BIDS claims that the owner of a naturally ventilated or mixed-mode building will achieve a 47–79 percent energy savings on HVAC systems, a 0.8–1.3 percent savings in health costs, and a 3–18 percent gain in productivity. They further calculate that this equates to an average 120 percent return on initial investment in one year.

Perhaps the biggest challenge that natural ventilation faces is overcoming entrenched attitudes in both the building industry and among building occupants. Over the past 50 years or more the use of air-conditioning has become the norm in modern buildings and carries a quality status associated with its use. Overcoming this perceived quality barrier may be the single biggest hurdle that the true adoption of 100 percent natural ventilation in tall office buildings faces. Even though the energy reductions and cost benefits are considerable, there is still a perceived quality and reliability issue with natural ventilation that pushes even the most environmentally conscious client toward adopting a hybrid ventilation system – with mechanical systems as a back-up to natural ventilation in the case of problems.

As mentioned at the outset of this technical guide, the true potential of delivering ventilation by natural means will only be achieved through 100 percent reliance on natural ventilation and the consequential removal of mechanical plant. Hybrid buildings, by still providing back-up mechanical equipment (sized predictably to cope with peak climatic extremes/worst-case scenarios) reduce operating energy for a portion of the year, but do not eliminate the embodied energy that resides in the mechanical plant, nor the space needed to house such equipment. It may be some time until systems of natural ventilation progress to the point where clients feel confident in relying exclusively on them. It is only at that point when the industry will be able to say that the true potential of natural ventilation has been delivered.

**Hybrid buildings reduce operating energy for a portion of the year, but do not eliminate the embodied energy that resides in the mechanical plant, nor the space needed to house such equipment.**

# 4.0 Recommendations and Future Research

# 4.0 Recommendations and Future Research

## 4.1 Recommendations

On the basis of the considerations presented in the previous section, the following recommendations for the natural ventilation of tall office buildings have been formulated:

### Thermal Comfort Standards:

▶ Utilize "adaptive" thermal comfort standards in determining the range of indoor comfort temperatures and periods during which the building can be naturally ventilated.

▶ Allow occupants a high degree of control over their direct environment and the operation of windows, while ensuring that this does not negatively affect the operation of the system overall, or adversely affect other occupiers.

▶ Consider the culture of the workplace and how the decision to changeover between natural ventilation and air-conditioning will take place (e.g., automatically, democratically, or autocratically).

### Local Climate:

▶ Study all aspects of local climate extensively, including consideration of how the direct urban microclimate around the site can affect directional wind speeds, temperatures, humidity, and day/night/seasonal variations. Do not rely exclusively on published climate data, which can be too general. Wherever possible, supplement this data with field observations.

▶ In extreme climates, provide supplemental heating or cooling of the intake air.

▶ Anticipate the risk of condensation in humid climates, especially if radiant slabs are used.

▶ When there is a significant difference between day/night temperatures in warm climates, use night-time ventilation to carry away the heat built up during the day in the building's exposed mass, to cool the building and reduce the cooling load the next day. Embrace this strategy further by creating larger areas of exposed thermal mass (e.g., exposed concrete ceiling slabs).

▶ In cold climates, consider the employment of glass double-skin façades, atria and/or winter gardens to preheat fresh, cold air before it enters the occupied office spaces.

▶ In cold climates, the use of a water feature within the spaces receiving incoming air could be used to humidify fresh air for natural ventilation.

▶ In a hot, arid environment, consider the interference of air-borne particulates (especially sand) in the natural ventilation system. Air-borne dust can be a problem with natural ventilation, even in temperate climates.

▶ Consider pest management at the lower levels of buildings, especially in tropical environments.

▶ Carefully factor in additional risk scenarios associated with future climate change. Aspects to consider include increasing urban heat island, wind strengths, heat waves, precipitation levels, and the frequency of severe weather events (non-typical periods of heat or cold) – all of which may be attenuated by climate change.

### Site Context, Building Orientation, and the Relative Driving Forces for Natural Ventilation:

▶ Orient the building and the air inlets toward the prevailing wind direction (particularly summer winds), especially if cross-ventilation is the primary natural ventilation strategy.

▶ Account for the effect of surrounding (and possible future) tall buildings on wind and airflow patterns around the building. The ideal site location would have minimal summer wind obstructions from the surroundings.

▶ Incorporate maximum daylighting in combination with natural ventilation to reduce artificial lighting and internal heat loads and increase natural ventilation effectiveness.

▶ Devise an appropriate solar control strategy to minimize solar heat gains and reduce the cooling load.

▶ Optimal cross-ventilation performance can be achieved by using the building shape and façade elements to generate a greater pressure differential on the windward and leeward faces of the building.

▶ The dominance of wind versus buoyancy requires special considerations of the opening geometry, which becomes a governing parameter in the airflow distribution.

▶ Consider using stack effect (which is intrinsic to tall buildings due to their height) as a ventilation strategy. Consider the use of atria, sky gardens and other vertical shafts to attenuate this stack effect, but be wary of creating extreme stacks which are difficult to control. Be cautious also of side-effects such as drafts, high wind speeds, etc.

▶ Whenever possible, rely on both wind-induced and buoyancy-induced ventilation to increase the magnitude of the driving forces, enhance the effectiveness of natural ventilation, and to obtain more consistent, stable ventilation rates.

### Planning and Spatial Configuration:

▶ Design tall buildings with narrow plan depths to facilitate the flow of air across interior spaces and enhance the effectiveness of natural ventilation.

▶ Locate open-plan office spaces along the perimeter of the building and locate enclosed rooms, meeting spaces and cellular offices toward the center.

▶ Alternatively, if cellular offices are located at the perimeter of the building, devise ways (e.g., "loose" partitions) to allow cross-ventilating air to move into the central spaces.

▶ Consider the placement of vertical circulation elements and service facilities where they don't obstruct the flow of air across the interior. Horizontal circulation elements can be used positively for airflow.

▶ Consider the exploitation of the stack effect in secondary, vertical circulation elements such as staircases and escalator voids, as long as they are not a means of egress.

▶ Consider interior design/layout within the greater ventilation strategy as internal partitions and office furniture will have a direct influence on the performance of natural ventilation.

▶ If night-time ventilation is employed, upstand structural beams (as opposed to the standard downstand beams) allow for a flat ceiling to be exposed as thermal mass, enabling the cool night-time air to pass directly against the slab and help remove daytime heat gains. Alternatively, downstand beams increase the surface area of thermal mass exposed to the circulating air and thus potentially increase the cooling, but then the beams need to be oriented in the direction of airflow so as not to impede ventilation.

### Sky Gardens and Vertical Segmentation of Atria:

▶ The sun's heat passing into an atrium or sky garden can be used to induce stack effect to assist with natural ventilation. Additionally, heat can be recovered from the exhaust air and used to preheat incoming air in the winter.

▶ Consider incorporating a central atrium (or open court) as an extraction chimney for exhaust air that induces the flow of air through the surrounding spaces.

▶ For larger floor plan depths, a central atrium can be used to directly ventilate interior-facing offices.

▶ Vertical segmentation should be utilized when an atrium or shaft runs the full height of the building to avoid the development

of extreme stack flows (and for greater smoke control).

▶ Coordinate and integrate vertical segmentation with the vertical transportation, (mixed-mode) HVAC zones/configurations, and fire spread control strategy.

### Aerodynamic Elements and Forms:

▶ Consider utilizing aerodynamic building forms that encourage the flow of wind around the exterior and into the building from a wide range of directions. Additionally, this can create a better environment for the pedestrian at street level.

▶ An opening in the building which acts as a "Wind Floor," or the placement of an aerodynamic element over a central stack such as a "Wing Roof," can provide additional uplift in the stack space and increase airflow rates in the occupied spaces.

▶ Aerodynamic elements such as "Wing Walls" can induce the flow of air in and out of the building, and capture winds from a wider range of directions.

### Façade Treatment and Double-Skin:

▶ The façade openings – whether in single or double-skin – should be oriented to achieve appropriate maximum ventilation levels.

▶ Size the exterior ventilation openings (and openings adjoining atria) to moderate various wind pressures across the façade. This can be achieved with the envelope flow model suggested in CIBSE AM10.

- Use blinds or perforated sun-screens (which are not glass) to give protection from direct solar heat gains and thus reduce internal heat loads on the ventilation system.

- Consider the use of glass double-skin façades for tall buildings that are subject to high wind speeds and external noise levels.

- Consider the use of glass double-skin façades in cold and temperate climates with a low relative humidity and low temperatures during the coldest months.

- Consider the use of "corridor" double-skin façades (with no vertical partitions) to balance varying wind forces and dissipate pressure differentials across the façade.

- Cavities which allow natural ventilation should be of sufficient width and need to be operable to provide access for cleaning and maintenance.

### Related Sustainable Strategies:

- Before taking specific actions toward a natural ventilation strategy, it is imperative to reduce heat loads overall, which is essential to the success of natural ventilation. This includes addressing broader issues of climate responsive and energy efficient design such as shading, thermal mass, daylighting, plug load reduction, etc.

### Predictive Performance and Modeling:

- Use CFD simulations and wind tunnel testing to predict airflow patterns, indoor air velocities, indoor temperature distribution, airflow rates, and thermal comfort associated with various outdoor wind speeds, weather conditions, and/or different combinations of window openings/configurations.

- CFD simulations can also inform the configuration and sizing of façade openings and to devise an appropriate control strategy for the operation of windows and vents.

- CFD simulations can test the effectiveness of natural ventilation in extreme temperatures and in "worst-case" areas of the building to ensure comfort is maintained in all zones during occupied hours.

- Use wind tunnel measurements of cladding pressures to drive calculations of flow rates.

## 4.2 Future Research

As there are relatively few tall buildings in existence which are entirely naturally ventilated, due to the potential risks, challenges, and associated limitations, the subject matter is worthy of greater investigation and future research. The topics below are suggested as areas requiring further research for the advancement of natural ventilation in tall buildings, portrayed as potential stand-alone research projects:

### Determining Adaptive Thermal Comfort Standards

A global study – based on post-occupancy evaluation of actual buildings – to determine the range of thermal comfort standards that are acceptable in office spaces, and how these are influenced by culture, climate, occupant make-up, and other factors. The study should also embrace what other elements could contribute to more tolerance of a wider range of acceptable thermal comfort standards (e.g., clothing, working patterns, etc.).

### A Global Post-Occupancy Analysis

A comprehensive global post-occupancy study needs to be undertaken of a wide range of naturally ventilated buildings from different climates, cultures, building types, functions, etc. This analysis should embrace both the energy usage in the building (embodied as well as operational) and occupant feedback with regards to the experience of natural ventilation. Comparison to mechanically ventilated and mixed-mode buildings in the study is also important, as is the embrace of automatically or manually controlled systems.

### The true costs of implementing Natural Ventilation

A study should be conducted considering the full holistic financial implications of implementing natural ventilation over mechanical systems. The study should embrace construction and material costs, operating costs, considerations of floor space usage (e.g., loss of space through double-skin versus saving of space through no mechanical plant), productivity gains of employees through a healthier environment, etc.

### Impact on Building Codes and Guidelines

A study of the full impact of incorporating natural ventilation (adaptive thermal standards, removal of mechanical back-up, impact on fire and life safety, attitude toward double-skin façades, etc.) on differing codes and standards around the world. For example, while the use of a double-skin façade may be beneficial to certain projects, some building codes reduce this benefit by including it in FAR (floor area ratio) calculations. Further investigations should assess the appropriateness of including unoccupied spaces, such as double-skin façades, in zoning calculations.

### Advancing Predictive Analysis

Technological advances should make predictive performance analysis more accurate and easier / more cost effective to implement in the design stage. As the risks to our urban centers increases with advancing climate change, the ability to more accurately predict weather patterns and how buildings respond in differing scenarios is vitally important.

### Site/Climate-Specific Strategies

Further studies should be conducted to implement climate-specific natural ventilation strategies, especially in climates with extreme temperature and humidity levels. Climate/local-specific "bio-mimicry" approaches could provide research and insight in designing systems that adapt and benefit from sun, wind, temperature, and humidity patterns. While high-density urban environments pose a challenge to natural ventilation strategies, some sites benefit from induced wind speeds from urban corridors. Future planning research could determine the feasibility of an urban network cohesively working together so multiple buildings could benefit from natural ventilation.

### Aerodynamic Elements

Several case studies in this technical guide utilize aerodynamic elements to induce natural ventilation. Further exploration should be carried out on such elements that induce, mitigate, and direct incoming fresh air. Such examples include: the use of solar chimneys for unsteady wind conditions; wing walls to increase façade pressure differential and stagnant wind speeds; wing roofs to accelerate exhaust from thermal flues; detailed aerodynamic elements in window mullions to disperse ventilation, etc.

### Façades for Natural Ventilation

A comprehensive study of façades – both single and double-skin – and their impact on natural ventilation should be carried out. Greater consideration of façade materials can have an impact on thermal conductivities and heat storage capacities. Phase change materials in combination with natural ventilation can further reduce energy consumption and maintain thermal comfort conditions. Research on various solar shading strategies which do not limit airflow rates would be advantageous.

### Future Technologies

As future technologies incorporate heat-recovery systems prior to discharging stale air, other energy efficient strategies need also be explored. Further research should be conducted in areas such as the use of wind turbines in combination with natural ventilation systems, solar cooling, and the use of water features to humidify and/or cool incoming fresh air.

### Evaporative Cooling

Evaporative cooling is a strategy to cool the air through the vaporization of water and can be especially well suited for hot, dry climates with a low relative humidity. While the use of a water feature or fountain has been previously discussed in this guide, other strategies should be researched and suggested for use in commercial applications. Such research examples include the integration of evaporative cooling strategies within the façade (see also Phase Change Materials below).

### Phase Change Materials

When designing low-energy buildings, greater consideration should be given to the façade materials and the different thermal conductivities and heat storage capacities they possess. The use of phase change materials (PCMs), which can absorb heat during the day and release it during the night, can reduce the need for mechanical ventilation to offset internal heat gains in some climates. Future research should be conducted on the commercial application of PCMs and the longevity of such materials and their containers.

### Fire and Life Safety in the Context of Natural Ventilation

Risk assessment methodologies should be further developed and standardized to ensure that natural ventilation does not introduce a level of risk with fire and safety that would have not existed in a mechanically conditioned building. Several factors worth consideration include: the reliability of the automatic sprinkler system, building compartmentalization features, and building evacuation strategies.

### Pollutants and Other Air-borne Risks

Considerations should be given to indoor air pollutants and other air-borne risks, to determine strategies to mitigate their infiltration. This is especially true in an age of potentially increasing biological and chemical warfare or terrorism. Buildings that rely on natural ventilation are perhaps at greater risk to these exposures and research needs to be undertaken to understand the risks and possible remedies.

# 5.0 References

# Bibliography

**Books:**

Allard, F. & Santamouris, M. (1998) *Natural Ventilation in Buildings: A Design Handbook.* Routledge: New York.

Baird, G. (2001) *The Architectural Expression of Environmental Control Systems.* Spon Press: London.

Blaser, W. (2003) *Post Tower: Helmut Jahn, Werner Sobek, Matthias Schuler.* Birkhäuser: Basel.

Briegleb, T. (ed.) (2000) *Ingenhoven Overdiek and Partners: High-Rise RWE AG Essen.* Birkhäuser: Basel.

Carter, B. (2008) *GSA / Morphosis / Arup: Integrated Design – San Francisco Federal Building.* Buffalo Workshop: New York.

Castillo, P. (2001) "Two towers," in Beedle, L. (ed.) *Cities in the Third Millennium – Proceeding of the CTBUH 6th World Congress.* Spon Press: London, pp. 413–416.

Cziesielski, E. (ed.) (2003) *Bauphysik-Kalender 2003.* Ernst & Sohn: Berlin, pp. 633–646.

Davies, C. & Lambot, I. (1997) *Commerzbank Frankfurt: Prototype for an Ecological High-Rise.* Watermark Publications: Basel.

Eisele, J. & Kloft, E. (2003) *High-rise Manual: Typology and Design, Construction and Technology.* Birkhäuser: Basel, pp. 186–188.

Etheridge, D. & Ford, B. (2008) "Natural ventilation of tall buildings – options and limitations" in Wood, A. (ed.) *Tall & Green: Typology for a Sustainable Urban Future – Proceedings of the CTBUH 8th World Congress.* Synergy: Dubai, pp. 226–232.

Feireiss, K. (2002) *Ingenhoven Overdiek and Partners, Energies.* Birkhäuser: Basel, pp. 214–241.

Fischer, V. (1997) *Sir Norman Foster and Partners: Commerzbank, Frankfurt am Main.* Axel Menges: Stuttgart.

Gonçalves, J. (2010) *The Environmental Performance of Tall Buildings.* Earthscan Ltd.: London.

Herzog, T. (2000) *Sustainable Height: Deutsche Messe AG Hannover Administration Building.* Prestel Press: Munich.

Ingenhoven, C. (2001) "Greening office towers," in Beedle, L. (ed.) *Cities in the Third Millennium – Proceedings of the CTBUH 6th World Congress.* Spon Press: London, pp. 527–530.

Irving, S., Ford, B. & Etheridge, D. (2005) *Natural Ventilation in Non-Domestic Buildings.* The Chartered Institution of Building Services Engineers (CIBSE): London.

Ismail, L. H. & Sibley, M. (2006) "Bioclimatic performance of high-rise office buildings: a case study in Penang," in Compagnon, C., Haefeli, P. & Weber, W. (eds.) *Proceedings of PLEA 2006 23rd International Conference.* Imprimerie St-Paul: Fribourg, Switzerland, pp. 981–986.

Jahnkassim, P. S. & Ip, K. (2000) "Energy and occupant impacts of bioclimatic high-rises in a tropical climate," in Steemers, K. & Yannas, S. (eds.) *Proceedings of PLEA 17th International Conference.* James and James (Science Publishers) Ltd.: London, pp. 249–250.

Jenkins, D. (ed.) (2009) "Swiss Re Headquarters," in *Norman Foster. Works 5.* Prestel: Munich, pp. 488–531.

Jenkins, D. (ed.) (2004) *Norman Foster: Works 4.* Prestel: Munich, pp. 35–89.

Krauel, J. (ed.) (2008) "Murphy/Jahn: Highlight Munich Business Towers," in *Corporate Buildings.* Links Books: Barcelona, pp. 82–91.

Kuwabara, B., Auer, T., Gouldsborough, T., Akerstream, T. & Klym, G. (2009) "Manitoba Hydro Place: integrated design process exemplar," in Demers, C. & Potvin, A. (eds.) *Proceedings of PLEA 26th International Conference.* Les Presses de l'Université Laval: Quebec City, pp. 551–556.

Lakkas, T. & Mumovic, D. (2009) "Sustainable cooling strategies," in Mumovic, D. & Santamouris, M. (eds.) *A Handbook of Sustainable Building Design & Engineering: An Integrated Approach to Energy, Health and Operational Performance.* Earthscan: London, pp. 287–290.

McConahey, E., Haves, P. & Christ, T. (2002) "The integration of engineering and architecture: a perspective on natural ventilation for the new San Francisco Federal Building," in *2002 ACEEE Summer Study on Energy Efficiency in Buildings.* American Council for an Energy Efficient Economy: Washington, DC, pp. 239–252.

Moe, K. (2008) *Integrated Design in Contemporary Architecture.* Princeton Architectural Press: Princeton, pp. 18–23.

Oesterle, E., Lieb, R., Lutz, M. & Heusler, W. (2001) *Double-Skin Façades: Integrated Planning.* Prestel: Munich.

Pomeroy, J. (2008) "Sky courts as transitional space: using space syntax as a predictive theory," in Wood, A. (ed.) *Tall & Green: Typology for a Sustainable Urban Future – Proceedings of the CTBUH 8th World Congress.* Synergy: Dubai, pp. 580–587.

Powell, K. (2006) *30 St Mary Axe: A Tower for London.* Merrell: London.

Powell, R. (1999) *Rethinking the Skyscraper: The Complete Architecture of Ken Yeang.* Thames and Hudson: London, pp. 82–91.

Quantrill, M. (1999) *The Norman Foster Studio: Consistency through Diversity.* E & FN Spon: London, pp. 164–169.

Richards, I. (2001) *T. R. Hamzah and Yeang: Ecology of the Sky.* The Images Publishing Group: Mulgrave, Australia, pp. 170–181.

Sauerbruch, M. & Hutton, L. (eds.) (2000) *GSW Headquarters, Berlin. Sauerbruch Hutton Architects.* Lars Müller Publishers: Baden.

Schmidt, C. (2006) *Highlight Towers.* Braun Publishing: Berlin.

Wells, M. (2005) *Skyscrapers: Structure and Design.* Laurence King Publishing Ltd.: London, pp. 86–91.

Wigginton, M. & Harris, J. (2004) *Intelligent Skins.* Architectural Press: Oxford, pp. 49–54.

Wood, A. (ed.) (2011) *Best Tall Buildings 2011: CTBUH International Award Winning Projects.* Routledge: New York, pp. 118–123.

Wood, A. (ed.) (2010) *Best Tall Buildings 2009: CTBUH International Award Winning Projects.* Routledge: New York, pp. 20–27.

Wood, A. (ed.) (2008) *Best Tall Buildings 2008: CTBUH International Award Winning Projects.* Elsevier Inc./ Architectural Press: Burlington, pp. 30–31.

Yeang, K. (2008) "Ecoskyscrapers and ecomimesis: new tall building typologies," in Wood, A. (ed.) *Tall & Green: Typology for a Sustainable Urban Future – Proceedings of the CTBUH 8th World Congress.* Synergy: Dubai, pp. 84–94.

Yeang, K. (2006) *Ecodesign: A Manual for Ecological Design.* Wiley-Academy: Chichester.

**Journal Articles:**

Anna, S. (2007) "Highlight Business Towers, Munich, Germany, 2004," *World Architecture (China),* vol. 210, pp. 50–57.

Arnold, D. (1999a) "The evolution of modern office buildings and air conditioning," *ASHRAE Journal,* vol. 41, no. 6, pp. 40–54.

Arnold, D. (1999b) "Air conditioning in office buildings after World War II," *ASHRAE Journal,* vol. 41, no. 7, pp. 33–41.

Arriola Clemenz, S. & Pérez-Torres, A. (2006) "Torre Cube, Puerta de Hierro, Guadalajara, Mexico," *On Diseño,* vol. 47, no. 9, pp. 86–107.

Bailey, P. (1997) "Commerzbank, Frankfurt; architects: Foster + Partners," *Arup Journal,* vol. 32, no. 2, pp. 3–12.

Brager, G. S. (2006) "Mixed-mode cooling," *ASHRAE Journal,* vol. 48, no. 8, pp. 30–37.

Brager, G. S. & de Dear, R. (2002) "Climate, comfort, & natural ventilation: a new adaptive comfort standard for ASHRAE Standard 55," *Energy and Buildings,* vol. 34, no. 6, pp. 549–561.

Brager, G. S. & de Dear, R. (2000) "A standard for natural ventilation," *ASHRAE Journal,* vol. 42, no. 10, pp. 21–28.

Brager, G. S., Ring, E. & Powell, K. (2000) "Mixed-mode ventilation: HVAC meets mother nature," *Engineered Systems,* May, pp. 60–70.

Buccino, G. (2004) "Deutsche Post, Bonn," *Industria Delle Construzioni,* vol. 38, no. 376, pp. 64–71.

Carrilho da Graça, G., Linden, P. F. & Haves, P. (2004) "Design and testing of a control strategy for a large, naturally ventilated office building," *Building Services Engineering Research and Technology Journal,* vol. 25, no. 3, pp. 223–239.

Chodikoff, I. (2006) "Award of Excellence: Manitoba Hydro Head Office," *Canadian Architect,* vol. 51, no. 12, pp. 32–35.

Chye, L.P. (1998) "Menara UMNO, Jalan Macalister, Pulau Pinang," *Architecture Malaysia,* vol. 10, no. 1, pp. 32–35.

Clemmetsen, N., Muller, W. & Trott, C. (2000) "GSW Headquarters, Berlin," *Arup Journal,* vol. 35, no. 2, pp. 8–12.

Dassler, F. H. (2007) "Extreme Randbedungen: Interview with Bruce Kuwabara and Thomas Auer," *XIA Intelligente Architetektur,* vol. 58, no. 1, pp. 18–25.

Dassler, F. H., Sobek, W., Reuss, S. & Schuler, M. (2003) "Post Tower in Bonn: state of the art," *XIA Intelligente Architetektur,* vol. 41, pp. 22–37.

Evans, B. (1997) "Through the glass cylinder," *Architect's Journal,* vol. 205, no. 19, pp. 42–45.

Ford, B. (2000) "Herzog in Hannover," *Ecotech,* vol. 1, pp. 4–7.

Galiano, L. F. (2005) "Torre Cube, Guadalajara (Mexico)," *AV Monografías,* vol. 115, pp. 18–24.

Gonçalves, J. & Bode, K. (2011) "The importance of real life data to support environmental claims for tall buildings," *CTBUH Journal,* vol. 2, pp. 24–29.

Gonçalves, J. & Bode, K. (2010) "Up in the air," *CIBSE Journal,* December, pp. 32–34.

Gonchar, J. (2010) "More than skin deep," *Architectural Record,* vol. 198, no. 7, pp. 102–110.

Gregory, R. (2004) "Squaring the circle," *Architectural Review,* vol. 215, no. 1287, pp. 80–85.

Haves, P., Linden, P. F. & Carrilho da Graça, G. (2004) "Use of simulation in the design of a large naturally venti-lated office building," *Building Services Engineering Research and Technology Journal,* vol. 25, no. 3, pp. 211–221.

Hodder, S. (2001) "GSW Headquarters, Berlin; architects: Sauerbruch Hutton," *Architecture Today,* vol. 116, pp. 30–49.

Kolokotroni, M. & Aronis, A. (1999) "Cooling-energy reduction in air-conditioned offices using night ventilation," *Applied Energy,* vol. 63, no. 4, pp. 241–253.

Lehmann, S. & Ingenhoven, C. (2009) "The future is green – a conversation between two German architects in Sydney," *Journal of Green Building,* vol. 4, no. 3, pp. 44–51.

Linn, C. (2011) "Architects: Architectus and Ingenhoven Architects," *Architect,* vol. 100, no. 11, pp. 84–88.

Linn, C. (2010) "Cold comfort," *GreenSource,* March/April, pp. 52–57.

Lu, W. (2007) "Special issue. Pioneering sustainable design: Thomas Herzog + Partner," *World Architecture (China),* vol. 204, pp. 16–79.

Metz, T. (2006) "Building types study: 855. Offices: culture clash: Highlight Munich Business Towers," *Architectural Record,* vol. 194, no. 3, pp. 154–160.

Meyer, U. (2011) "Colors 'n curves: a new bank headquarters in Frankfurt may well be the world's most energy-efficient office tower – KFW Westarkade, Frankfurt, Germany," *GreenSource,* May/June, pp. 48–53.

Meyer, U. (2008) "Double-skin deep: designing environmentally sustainable architecture often depending on the façade," *World Architecture (China),* vol. 214, pp. 18–25.

Pasquay, T. (2004) "Natural ventilation in high-rise buildings with double façades, saving or waste of energy," *Energy and Buildings,* vol. 36, no. 4, pp. 381–389.

Pearson, J. (1997) "Delicate Essen," *Architectural Review,* vol. 202, no. 1205, pp. 40–45.

Pepchinski, M. (2002) "GSW Headquarters, Berlin," *A&U: Architecture and Urbanism,* vol. 384, no. 9, pp. 64–73.

Pepchinski, M. (1998) "With its naturally ventilated skin and gardens in the sky, Foster + Partners" Commerzbank reinvents the skyscraper," *Architectural Record,* vol. 186, no. 1, pp. 68–79.

Pepchinski, M. (1997) "RWE AG Hochhaus – Essen, Germany," *Architectural Record,* vol. 185, No. 6, pp. 144–151.

Powell, R. (1998) "Vertical aspirations – Menara UMNO: Penang, Malaysia; architects: T. R. Hamzah & Yeang," *Singapore Architect,* vol. 200, pp. 66–71.

Russell, J. (2000) "GSW Headquarters, Berlin," *Architectural Record,* vol. 188, no. 6, pp. 156–164.

Sampson, P. (2010) "Climate-controlled," *Canadian Architect,* vol. 55, no. 1, pp. 16–22.

Sauerbruch, M. (2011) "Sustainable architecture," *Detail Green English,* vol. 1, pp. 26–31.

Sauerbruch, M. & Hutton, L. (2001) "GSW Headquarters, Berlin, Germany; Sauerbruch Hutton Architects," *UME,* vol. 13, pp. 24–37.

Schittich, C. (ed.) (2007) "Torre Cube in Guadalajara, Mexico," *Detail,* vol. 47, no. 9, pp. 962–964, 1076.

Schuler, M. (2004) "Interview with Matthias Schuler: Sustainable coopera-tion nowadays – interaction between architects and climate engineers," *A&U: Architecture and Urbanism,* vol. 410, no. 11, pp. 112–121.

Seguin, B. (2011) "Façade technology," *Architecture Australia,* vol. 100, no. 3, pp. 104–105.

Slavic, D. (2008) "IBS Award 2008," *XIA International Magazine,* vol. 8, no. 2, pp. 11–17.

Spring, M. (2008) "Projects. Gherkin revisited; architects: Foster + Partners," *Building,* vol. 273, no. 8526(16), pp. 62–67.

Thierfelder, A. & Schuler, M. (2009) "In site: site specificity in sustainable architecture," *Harvard Design Magazine,* vol. 30, pp. 50–59, 153.

Vivian, P. (2008) "Space: next-generation workspace," *Architecture Australia,* vol. 97, no. 3, pp. 97–102.

**Conference Papers:**

Chikamoto, T., Kato, S. & Ikaga, T. (1999) "Hybrid air-conditioning system at Liberty Tower of Meiji University," paper presented at 1999 IEA Energy Conservation in Buildings & Community Systems Annex 35 Conference on Hybrid Ventilation, Sydney, Australia, 28 September–1 October.

Jahnkassim, P. S. & Ip, K. (2006) "Linking bioclimatic theory and environmental performance in its climatic and cultural context – an analysis into the tropical high-rises of Ken Yeang," paper presented at PLEA 2006 Conference on Passive and Low Energy Architecture, Geneva, Switzerland, 6–8 September.

Jones, P. J. & Yeang, K. (1999) "The use of the wind wing-wall as a device for low-energy passive comfort cooling in a high-rise tower in the warm mid tropics," paper presented at PLEA 16th International Conference, Brisbane, Australia, 22–24 September.

Kato, S. & Chikamoto, T. (2002) "Pilot study report: the Liberty Tower of Meiji University," paper presented at 2002 IEA Energy Conservation in Buildings & Community Systems Annex 35 Conference on Hybrid Ventilation, Montreal, Canada, 13–17 May.

**Reports:**

Auliciems, A. & Szokolay, S. (1997) *Thermal comfort*. PLEA in association with Dept. of Architecture, University of Queensland: Brisbane, p. 14.

Bibb, D. (ed.) (2008) "The case for sustainability," in *Sustainability Matters*. US General Services Administration: Washington, DC, pp. 8–33.

Brager, G., Borgeson, S. & Lee, Y. (2007) *Summary Report: Control Strategies for Mixed-Mode Buildings*. Center for the Built Environment (CBE), University of California, Berkeley, pp. 34–38.

Loftness, V., Hartkopf, V. & Gurtekin, B. (2004) "Building investment decision support (BIDS): cost–benefit tool to promote high performance components, flexible infrastructures and systems integration for sustainable commercial buildings and productive organizations," Carnegie Mellon University Center for Building Performance and Diagnostics.

**Theses:**

Arons, D. M. M. & Glicksman L. R. (2008) "Double-skin, airflow façades: will the popular European model work in the USA?," PhD Thesis, Masachusetts Institute of Technology.

Cook, M. (1998) "An evaluation of computational fluid dynamics for modeling buoyancy-driven displacement ventilation," PhD Thesis, De Montfort University.

Jankassim, P. S. (2004) "The bioclimatic skyscraper: a critical analysis of the theories and designs of Ken Yeang," PhD thesis, University of Brighton.

Kleiven, T. (2003) "Natural ventilation in buildings: architectural concepts, consequences, and possibilities," PhD thesis, Norwegian University of Science and Technology.

# 100 Tallest Buildings in the World (as of April 2012)

The Council maintains the official list of the 100 Tallest Buildings in the World, which are ranked based on the height to architectural top, and includes not only completed buildings, but also buildings currently under construction. However, a building does not receive an official ranking number until it is completed (see criteria, pages 178–181).

*Color Key:*

Buildings listed in black are completed and officially ranked by the CTBUH.

Buildings listed in green are under construction and have topped out.

Buildings listed in red are under construction, but have not yet topped out.

| Rank | Building Name | City | Year | Stories | Height m | Height ft | Material | Use |
|------|---------------|------|------|---------|----------|-----------|----------|-----|
| 1 | Burj Khalifa | Dubai | 2010 | 163 | 828 | 2717 | steel/concrete | office/residential/hotel |
| | Ping An Finance Center | Shenzhen | 2015 | 115 | 660 | 2165 | composite | office |
| | Shanghai Tower | Shanghai | 2014 | 121 | 632 | 2073 | composite | hotel/office |
| | Makkah Royal Clock Tower Hotel | Mecca | 2012 | 120 | 601 | 1972 | steel/concrete | other/hotel/multiple |
| | Goldin Finance 117 | Tianjin | 2015 | 117 | 597 | 1957 | composite | hotel/office |
| | Lotte World Tower | Seoul | 2015 | 123 | 555 | 1819 | composite | hotel/office |
| | One World Trade Center | New York | 2013 | 104 | 541 | 1776 | composite | office |
| | The CTF Guangzhou | Guangzhou | 2017 | 111 | 530 | 1739 | composite | hotel/residential/office |
| | Dalian Greenland Center | Dalian | 2016 | 88 | 518 | 1699 | composite | hotel/residential/office |
| | Busan Lotte World Tower | Busan | 2016 | 107 | 510 | 1674 | composite | residential/hotel/office |
| 2 | Taipei 101 | Taipei | 2004 | 101 | 508 | 1667 | composite | office |
| 3 | Shanghai World Financial Center | Shanghai | 2008 | 101 | 492 | 1614 | composite | hotel/office |
| 4 | International Commerce Centre | Hong Kong | 2010 | 108 | 484 | 1588 | composite | hotel/office |
| | International Commerce Center 1 | Chongqing | 2016 | 99 | 468 | 1535 | composite | hotel/office |
| | Tianjin R&F Guangdong Tower | Tianjin | 2015 | 91 | 468 | 1535 | composite | residential/hotel/office |
| 5 | Petronas Tower 1 | Kuala Lumpur | 1998 | 88 | 452 | 1483 | composite | office |
| 5 | Petronas Tower 2 | Kuala Lumpur | 1998 | 88 | 452 | 1483 | composite | office |
| 7 | Zifeng Tower | Nanjing | 2010 | 66 | 450 | 1476 | composite | hotel/office |
| 8 | Willis Tower | Chicago | 1974 | 108 | 442 | 1451 | steel | office |
| | World One | Mumbai | 2015 | 117 | 442 | 1450 | composite | residential |
| 9 | Kingkey 100 | Shenzhen | 2011 | 100 | 442 | 1449 | composite | hotel/office |
| 10 | Guangzhou International Finance Center | Guangzhou | 2010 | 103 | 439 | 1439 | composite | hotel/office |
| | Wuhan Center | Wuhan | 2015 | 88 | 438 | 1437 | - | hotel/residential/office |
| | Diamond Tower | Jeddah | - | 93 | 432 | 1417 | - | residential |
| 11 | Trump International Hotel & Tower | Chicago | 2009 | 98 | 423 | 1389 | concrete | residential/hotel |
| 12 | Jin Mao Building | Shanghai | 1999 | 88 | 421 | 1380 | composite | hotel/office |
| | Princess Tower | Dubai | 2012 | 101 | 413 | 1356 | steel/concrete | residential |
| 13 | Al Hamra Firdous Tower | Kuwait City | 2011 | 77 | 413 | 1354 | concrete | office |
| | Marina 101 | Dubai | 2012 | 101 | 412 | 1352 | concrete | residential/hotel |
| 14 | Two International Finance Centre | Hong Kong | 2003 | 88 | 412 | 1352 | composite | office |
| | Two World Trade Center | New York | - | 79 | 411 | 1348 | composite | office |
| 15 | 23 Marina | Dubai | 2011 | 90 | 393 | 1289 | concrete | residential |
| 16 | CITIC Plaza | Guangzhou | 1996 | 80 | 390 | 1280 | concrete | office |
| | Capital Market Authority Headquarters | Riyadh | 2013 | 77 | 385 | 1263 | composite | office |
| | Forum 66 Tower 1 | Shenyang | 2014 | 76 | 384 | 1260 | steel | office |
| 17 | Shun Hing Square | Shenzhen | 1996 | 69 | 384 | 1260 | composite | office |
| | Eton Place Dalian Tower 1 | Dalian | 2013 | 80 | 383 | 1257 | composite | hotel/office |
| | Abu Dhabi Plaza | Astana | 2016 | 88 | 382 | 1253 | - | residential |
| | The Domain | Abu Dhabi | 2013 | 88 | 381 | 1251 | concrete | residential |
| 18 | Empire State Building | New York | 1931 | 102 | 381 | 1250 | steel | office |
| | Elite Residence | Dubai | 2012 | 87 | 380 | 1248 | concrete | residential |
| | Three World Trade Center | New York | - | 71 | 378 | 1240 | composite | office |
| 19 | Central Plaza | Hong Kong | 1992 | 78 | 374 | 1227 | concrete | office |
| | Oberoi Oasis Residential Tower | Mumbai | - | 82 | 372 | 1220 | - | residential |
| | Zhujiang New City Tower | Guangzhou | - | 87 | 371 | 1217 | - | office |
| 20 | Bank of China Tower | Hong Kong | 1990 | 72 | 367 | 1205 | composite | office |
| 21 | Bank of America Tower | New York | 2009 | 55 | 366 | 1200 | composite | office |
| | VietinBank Business Center Office Tower | Hanoi | 2014 | 68 | 363 | 1191 | composite | office |
| | Federation Towers – Vostok Tower | Moscow | 2013 | 93 | 360 | 1181 | concrete | residential/hotel/office |
| 22 | Almas Tower | Dubai | 2008 | 68 | 360 | 1181 | concrete | office |

| Rank | Building Name | City | Year | Stories | Height m | ft | Material | Use |
|------|---------------|------|------|---------|----------|-----|----------|-----|
| | The Pinnacle | Guangzhou | 2012 | 60 | 360 | 1181 | concrete | office |
| | Emirates Park Towers Hotel & Spa 1 | Dubai | 2012 | 77 | 355 | 1165 | concrete | hotel |
| | Emirates Park Towers Hotel & Spa 2 | Dubai | 2013 | 77 | 355 | 1165 | concrete | hotel |
| 23 | Emirates Tower One | Dubai | 2000 | 54 | 355 | 1163 | composite | office |
| | Forum 66 Tower 2 | Shenzhen | 2014 | 68 | 351 | 1150 | steel | office |
| | Lamar Tower 1 | Jeddah | 2013 | 72 | 350 | 1148 | concrete | residential/office |
| 24 | Tuntex Sky Tower | Kaohsiung | 1997 | 85 | 348 | 1140 | composite | hotel/office |
| 25 | Aon Center | Chicago | 1973 | 83 | 346 | 1136 | steel | office |
| 26 | The Center | Hong Kong | 1998 | 73 | 346 | 1135 | steel | office |
| 27 | John Hancock Center | Chicago | 1969 | 100 | 344 | 1128 | steel | residential/office |
| | ADNOC Headquarters | Abu Dhabi | 2013 | 76 | 342 | 1122 | concrete | office |
| | Ahmed Abdul Rahim Al Attar Tower | Dubai | 2013 | 76 | 342 | 1122 | concrete | residential |
| | The Wharf Times Square 1 | Wuxi | 2015 | 68 | 339 | 1112 | composite | hotel/office |
| | Chongqing World Financial Center | Chongqing | 2013 | 73 | 339 | 1112 | composite | office |
| | Orchid Crown Tower A | Mumbai | 2014 | 75 | 337 | 1106 | concrete | residential |
| | Orchid Crown Tower B | Mumbai | 2014 | 75 | 337 | 1106 | concrete | residential |
| | Orchid Crown Tower C | Mumbai | 2014 | 75 | 337 | 1106 | concrete | residential |
| 28 | Tianjin Global Financial Center | Tianjin | 2011 | 74 | 337 | 1105 | composite | office |
| 29 | The Torch | Dubai | 2011 | 79 | 337 | 1105 | concrete | residential |
| | Keangnam Hanoi Landmark Tower | Hanoi | 2012 | 70 | 336 | 1102 | concrete | hotel/residential/office |
| | 16a IBC Tower 1 | Moscow | - | 91 | 336 | 1101 | concrete | office |
| 30 | Shimao International Plaza | Shanghai | 2006 | 60 | 333 | 1094 | concrete | hotel/office |
| | South Asian Gate | Kunming | 2014 | 83 | 333 | 1093 | - | residential/hotel/office |
| | Tianjin Kerry Center | Tianjin | 2014 | 72 | 333 | 1093 | steel | office |
| 31 | Rose Rayhaan by Rotana | Dubai | 2007 | 71 | 333 | 1093 | composite | hotel |
| | Mercury City Tower | Moscow | 2012 | 70 | 332 | 1089 | concrete | residential/office |
| | Modern Media Center | Changzhou | 2013 | 57 | 332 | 1089 | composite | office |
| 32 | Minsheng Bank Building | Wuhan | 2008 | 68 | 331 | 1086 | steel | office |
| | Ryugyong Hotel | Pyongyang | 2012 | 105 | 330 | 1083 | concrete | hotel/office |
| | Gate of Kuwait Tower | Kuwait City | - | 80 | 330 | 1083 | concrete | hotel/office |
| 33 | China World Tower | Beijing | 2010 | 74 | 330 | 1083 | composite | hotel/office |
| | The Skyscraper | Dubai | - | 66 | 330 | 1083 | - | office |
| | Suning Plaza Tower 1 | Zhenjiang | - | 77 | 330 | 1082 | - | - |
| | Han Kwok City Center | Shenzhen (CN) | - | 80 | 329 | 1081 | - | multiple |
| | Orchid Heights 1 | Mumbai (IN) | 2014 | 80 | 328 | 1076 | concrete | residential |
| | Orchid Heights 2 | Mumbai (IN) | 2014 | 80 | 328 | 1076 | concrete | residential |
| 34 | Longxi International Hotel | Jiangyin (CN) | 2011 | 74 | 328 | 1076 | composite | residential/hotel |
| | Al Yaqoub Tower | Dubai (AE) | 2012 | 69 | 328 | 1076 | concrete | residential/hotel |
| | Nanjing World Trade Center Tower 1 | Nanjing (CN) | 2015 | 69 | 328 | 1076 | - | hotel/office |
| | Wuxi Suning Plaza 1 | Wuxi | 2014 | 68 | 328 | 1076 | composite | hotel/office |
| 35 | The Index | Dubai | 2010 | 80 | 326 | 1070 | concrete | residential/office |
| 36 | The Landmark | Abu Dhabi | 2012 | 72 | 324 | 1063 | concrete | residential/office |
| | Deji Plaza Phase 2 | Nanjing | 2013 | 62 | 324 | 1063 | composite | office |
| | Yantai Shimao No. 1 The Harbour | Yantai | 2013 | 59 | 323 | 1060 | composite | residential/hotel/office |
| 37 | Q1 | Gold Coast | 2005 | 78 | 323 | 1058 | concrete | residential |
| | Gate of Taipei Tower 1 | Taipei | - | 76 | 322 | 1057 | - | hotel/office/retail |
| 38 | Wenzhou Trade Center | Wenzhou | 2011 | 68 | 322 | 1056 | concrete | hotel/office |
| 39 | Burj Al Arab Hotel | Dubai | 1999 | 60 | 321 | 1053 | composite | hotel |
| 40 | Nina Tower | Hong Kong | 2006 | 80 | 320 | 1051 | concrete | hotel/office |
| | Palais Royale | Mumbai | 2013 | 88 | 320 | 1050 | concrete | residential |
| | White Magnolia Plaza 1 | Shanghai | 2014 | 66 | 320 | 1048 | - | office |
| 41 | Chrysler Building | New York | 1930 | 77 | 319 | 1046 | steel | office |
| 42 | New York Times Tower | New York | 2007 | 52 | 319 | 1046 | steel | office |
| | Runhua Global Center 1 | Changzhou | 2014 | 72 | 318 | 1043 | - | office |
| | Riverside Century Plaza Main Tower | Wuhu | - | 66 | 318 | 1043 | - | hotel/office |
| 43 | HHHR Tower | Dubai | 2010 | 72 | 318 | 1042 | concrete | residential |
| 44 | Bank of America Plaza | Atlanta | 1993 | 55 | 317 | 1040 | composite | office |
| | Maha Nakhon | Bangkok | 2015 | 77 | 313 | 1028 | - | residential/hotel |
| | The Stratford Residences | Makati | 2015 | 70 | 312 | 1024 | - | residential |
| | Moi Center Tower A | Shenyang | 2012 | 75 | 311 | 1020 | composite | hotel/office |

| Rank | Building Name | City | Year | Stories | Height m | ft | Material | Use |
|------|---------------|------|------|---------|----------|-----|----------|-----|
| 45 | U.S. Bank Tower | Los Angeles | 1990 | 73 | 310 | 1018 | steel | office |
| 46 | Ocean Heights | Dubai | 2010 | 83 | 310 | 1017 | concrete | residential |
| 46 | Menara Telekom | Kuala Lumpur | 2001 | 55 | 310 | 1017 | concrete | office |
| | Pearl River Tower | Guangzhou | 2012 | 71 | 310 | 1016 | composite | office |
| | Guangzhou Fortune Center | Guangzhou | 2014 | 68 | 309 | 1015 | - | office |
| 48 | Emirates Tower Two | Dubai | 2000 | 56 | 309 | 1014 | concrete | hotel |
| | Shenyang New World Intl. Convention & Exhibition Center Tower 1 | Shenyang | 2014 | 60 | 308 | 1010 | - | office |
| | Shenyang New World Intl. Convention & Exhibition Center Tower 2 | Shenyang | 2014 | 60 | 308 | 1010 | - | office |
| | Lokhandwala Minerva | Mumbai | 2014 | 83 | 307 | 1007 | - | residential |
| 49 | Franklin Center – North Tower | Chicago | 1989 | 60 | 307 | 1007 | composite | office |
| | Infinity Tower | Dubai | 2013 | 76 | 306 | 1005 | concrete | residential |
| | One57 | New York | 2013 | 75 | 306 | 1005 | concrete | residential/hotel |
| | East Pacific Center Tower A | Shenzhen | 2013 | 85 | 306 | 1004 | concrete | residential |
| | The Shard | London | 2012 | 73 | 306 | 1004 | composite | residential/hotel/office |
| 50 | Etihad Tower T2 | Abu Dhabi | 2011 | 80 | 305 | 1002 | concrete | residential |
| 51 | JPMorgan Chase Tower | Houston | 1982 | 75 | 305 | 1002 | composite | office |
| 52 | Northeast Asia Trade Tower | Incheon | 2011 | 68 | 305 | 1001 | composite | residential/hotel/office |
| | Eurasia | Moscow | 2013 | 67 | 304 | 997 | composite | residential/hotel/office |
| 53 | Baiyoke Tower II | Bangkok | 1997 | 85 | 304 | 997 | concrete | hotel |
| | Shenzhen World Finance Center | Shenzhen | - | 68 | 304 | 997 | - | office |
| 54 | Two Prudential Plaza | Chicago | 1990 | 64 | 303 | 995 | concrete | office |
| | Diwang International Fortune Center | Liuzhou | 2014 | 75 | 303 | 994 | - | residential/hotel/office |
| | KAFD World Trade Center | Riyadh | 2014 | 67 | 303 | 994 | - | office |
| | Four Seasons Tower | Tianjin | 2014 | 65 | 303 | 994 | - | residential/hotel |
| | Leatop Plaza | Guangzhou | 2012 | 64 | 303 | 993 | composite | office |
| 55 | Wells Fargo Plaza | Houston | 1983 | 71 | 302 | 992 | steel | office |
| 56 | Kingdom Centre | Riyadh | 2002 | 41 | 302 | 992 | steel/concrete | residential/hotel/office |
| 57 | The Address | Dubai | 2008 | 63 | 302 | 991 | concrete | residential/hotel |
| 58 | Capital City Moscow Tower | Moscow | 2010 | 76 | 302 | 990 | concrete | residential |
| | Gate of the Orient | Suzhou | 2013 | 68 | 302 | 990 | composite | residential/hotel/office |
| 59 | Doosan Haeundae We've the Zenith Tower A | Busan | 2011 | 80 | 301 | 988 | concrete | residential |
| | Lamar Tower 2 | Jeddah | 2013 | 62 | 301 | 988 | concrete | residential/office |
| | Heung Kong Tower | Shenzhen | 2013 | 70 | 301 | 987 | composite | hotel/office |
| | Dubai Pearl Tower | Dubai | 2014 | 73 | 300 | 984 | concrete | residential |
| | NBK Tower | Kuwait City | 2014 | 70 | 300 | 984 | concrete | office |
| | Jin Wan Plaza 1 | Tianjin | 2015 | 66 | 300 | 984 | - | hotel/office |
| | Gran Torre Costanera | Santiago | 2013 | 64 | 300 | 984 | concrete | office |
| | Namaste Tower | Mumbai | 2014 | 62 | 300 ** | 984 | - | hotel/office |
| 60 | Arraya Tower | Kuwait City | 2009 | 60 | 300 | 984 | concrete | office |
| | Abeno Harukas | Osaka | 2014 | 59 | 300 | 984 | steel | hotel/office/retail |
| | Shenglong Global Center | Fuzhou | - | 57 | 300 | 984 | - | office |
| 61 | Aspire Tower | Doha | 2007 | 36 | 300 | 984 | composite | hotel/office |
| | Langham Hotel Tower | Dalian | 2015 | 74 | 300 | 983 | - | residential/hotel |
| 62 | One Island East Centre | Hong Kong | 2008 | 69 | 298 | 979 | concrete | office |
| 63 | First Bank Tower | Toronto | 1975 | 72 | 298 | 978 | steel | office |
| | Yujiabao Administrative Cervices Center | Tianjin | - | 60 | 298 | 978 | - | office |
| | Four World Trade Center | New York | 2013 | 64 | 298 | 977 | composite | office |
| 64 | Eureka Tower | Melbourne | 2006 | 91 | 297 | 975 | concrete | residential |
| 65 | Comcast Center | Philadelphia | 2008 | 57 | 297 | 974 | composite | office |
| 66 | Landmark Tower | Yokohama | 1993 | 73 | 296 | 972 | steel | hotel/office |
| 67 | Emirates Crown | Dubai | 2008 | 63 | 296 | 971 | concrete | residential |
| | Xiamen Shimao Cross-Strait Plaza Tower B | Xiamen | 2015 | 67 | 295 | 969 | - | office |
| 68 | Khalid Al Attar Tower 2 | Dubai | 2011 | 66 | 294 | 965 | concrete | residential/office |
| 69 | Trump Ocean Club International Hotel & Tower | Panama City | 2011 | 68 | 293 | 961 | concrete | residential/hotel |
| 70 | 311 South Wacker Drive | Chicago | 1990 | 65 | 293 | 961 | concrete | office |
| | Greenland Puli Center | Jinan | 2015 | 61 | 293 | 960 | composite | residential/office |
| 71 | Sky Tower | Abu Dhabi | 2010 | 74 | 292 | 959 | concrete | residential/office |
| 72 | Haeundae I Park Marina Tower 2 | Busan | 2011 | 72 | 292 | 958 | composite | residential |

| Rank | Building Name | City | Year | Stories | Height m | ft | Material | Use |
|---|---|---|---|---|---|---|---|---|
| | Wuxi Maoye City – Marriott Hotel | Wuxi | 2013 | 72 | 292 | 958 | composite | hotel |
| 73 | SEG Plaza | Shenzhen | 2000 | 71 | 292 | 957 | concrete | hotel/office |
| 74 | 70 Pine Street | New York | 1932 | 67 | 290 | 952 | steel | office |
| | Powerlong Center Tower 1 | Tianjin | - | 59 | 290 | 951 | - | office |
| | Global Trade Plaza | Dongguan | 2012 | 68 | 289 | 948 | composite | hotel/office |
| | Busan Int. Finance Center Landmark Tower | Busan | 2014 | 63 | 289 | 948 | - | office |
| | Jiangxi Nanchang Greenland Central Plaza 1 | Nanchang | 2014 | 59 | 289 | 948 | - | office |
| | Jiangxi Nanchang Greenland Central Plaza 2 | Nanchang | 2014 | 59 | 289 | 948 | - | office |
| 75 | Key Tower | Cleveland | 1991 | 57 | 289 | 947 | composite | office |
| | Park Hyatt Guangzhou | Guangzhou | 2013 | 66 | 289 | 947 | composite | residential/hotel/office |
| 76 | Plaza 66 | Shanghai | 2001 | 66 | 288 | 945 | concrete | office |
| 77 | One Liberty Place | Philadelphia | 1987 | 61 | 288 | 945 | steel | office |
| | Soochow International Plaza East Tower | Huzhou | - | - | 288 | 945 | - | hotel/office |
| | Soochow International Plaza West Tower | Huzhou | - | - | 288 | 945 | - | residential |
| | Chongqing Poly Tower | Chongqing | 2012 | 58 | 287 | 941 | concrete | office/hotel |
| 78 | Millennium Tower | Dubai | 2006 | 59 | 285 | 935 | concrete | residential |
| 79 | Sulafa Tower | Dubai | 2010 | 75 | 285 | 935 | concrete | residential |
| | Yingli Tower | Chongqing | 2012 | 60 | 285 | 935 | concrete | office |
| 80 | Tomorrow Square | Shanghai | 2003 | 60 | 285 | 934 | concrete | residential/hotel/office |
| 81 | Columbia Center | Seattle | 1984 | 76 | 284 | 933 | composite | office |
| | D1 Tower | Dubai | 2013 | 80 | 284 | 932 | concrete | residential |
| | Three International Finance Center | Seoul | 2012 | 55 | 284 | 932 | composite | office |
| 82 | Chongqing World Trade Center | Chongqing | 2005 | 60 | 283 | 929 | concrete | office |
| 83 | Cheung Kong Centre | Hong Kong | 1999 | 63 | 283 | 928 | steel | office |
| 84 | The Trump Building | New York | 1930 | 71 | 283 | 927 | steel | office |
| | Al Hekma Tower | Dubai | 2013 | 64 | 282 | 925 | steel/concrete | office |
| 85 | Suzhou RunHua Global Building A | Suzhou | 2010 | 49 | 282 | 925 | composite | office |
| 86 | Doosan Haeundae We've the Zenith Tower B | Busan | 2011 | 75 | 282 | 924 | concrete | residential |
| | Trump International Hotel & Tower | Toronto | 2012 | 59 | 281 | 922 | concrete | residential/hotel |
| | Torre Vitri | Panama City | 2012 | 75 | 281 | 921 | concrete | residential |
| 87 | Bank of America Plaza | Dallas | 1985 | 72 | 281 | 921 | composite | office |
| 88 | Marina Pinnacle | Dubai | 2011 | 73 | 280* | 919 | concrete | residential |
| 88 | United Overseas Bank Plaza One | Singapore | 1992 | 66 | 280 | 919 | steel | office |
| 88 | Overseas Union Bank Centre | Singapore | 1986 | 63 | 280 | 919 | steel | office |
| 88 | Excellence Century Plaza Tower 1 | Shenzhen | 2010 | 60 | 280 | 919 | composite | office |
| 88 | Shimao Didang New City Main Tower | Shzoxing | 2012 | 56 | 280 | 919 | composite | hotel/office |
| | Eton Place Dalian Tower 2 | Dalian | 2013 | 62 | 279 | 917 | composite | residential/office |
| | Zhengzhou Greenland Plaza | Zhengzhou | 2012 | 56 | 279 | 916 | composite | hotel/office |
| 93 | Citigroup Center | New York | 1977 | 59 | 279 | 915 | steel | office |
| 94 | Hong Kong New World Tower | Shanghai | 2004 | 59 | 278 | 913 | composite | hotel/office/retail |
| | Trust Tower | Abu Dhabi | 2012 | 60 | 278 | 912 | concrete | office |
| 95 | Etihad Tower 1 | Abu Dhabi | 2011 | 69 | 278 | 911 | concrete | hotel/residential |
| 96 | Republic Plaza | Singapore | 1996 | 66 | 276 | 906 | composite | office |
| 97 | Diwang International Commerce Center | Nanning | 2006 | 54 | 276 | 906 | concrete | hotel/office |
| 98 | Scotia Tower | Toronto | 1989 | 68 | 275 | 902 | composite | office |
| 99 | Williams Tower | Houston | 1982 | 64 | 275 | 901 | steel | office |
| | Ilham Baru Tower | Kuala Lumpur | 2014 | 58 | 274 | 899 | - | residential/office |
| 100 | Nantong Zhongnan International Plaza | Nantong | 2011 | 53 | 273 | 897 | steel/concrete | residential/office |

* estimated height
** minimum height

# CTBUH Height Criteria

The Council on Tall Buildings and Urban Habitat is the official arbiter of the criteria upon which tall building height is measured, and the title of "The World's (or Country's, or City's) Tallest Building" determined. The Council maintains an extensive set of definitions and criteria for measuring and classifying tall buildings which are the basis for the official "100 Tallest Buildings in the World" list (see pages 174–177).

## What is a Tall Building?

There is no absolute definition of what constitutes a "tall building." It is a building that exhibits some element of "tallness" in one or more of the following categories:

- ▶ **Height relative to context:** It is not just about height, but about the context in which it exists. Thus, whereas a 14-story building may not be considered a tall building in a high-rise city such as Chicago or Hong Kong, in a provincial European city or a suburb this may be distinctly taller than the urban norm.

- ▶ **Proportion:** Again, a tall building is not just about height but also about proportion. There are numerous buildings which are not particularly high, but are slender enough to give the appearance of a tall building, especially against low urban backgrounds. Conversely, there are numerous big/large footprint buildings which are quite tall but their size/floor area rules them out as being classed as a tall building.

- ▶ **Tall Building Technologies:** If a building contains technologies which may be attributed as being a product of "tall" (e.g., specific vertical transport technologies, structural wind bracing as a product of height, etc.), then this building can be classed as a tall building.

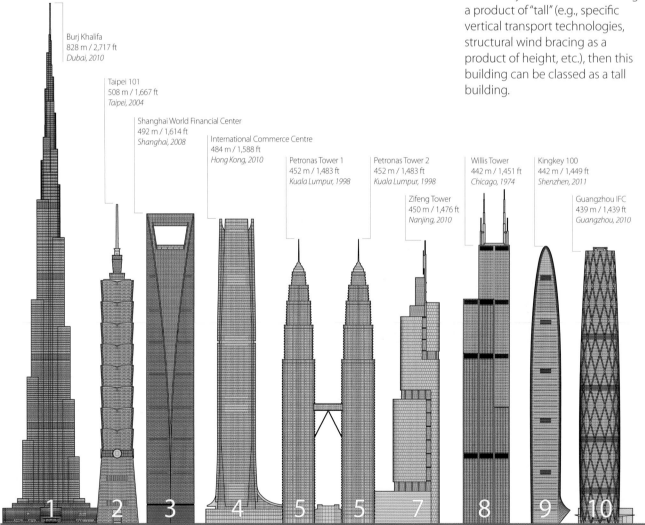

Burj Khalifa
828 m / 2,717 ft
*Dubai, 2010*

Taipei 101
508 m / 1,667 ft
*Taipei, 2004*

Shanghai World Financial Center
492 m / 1,614 ft
*Shanghai, 2008*

International Commerce Centre
484 m / 1,588 ft
*Hong Kong, 2010*

Petronas Tower 1
452 m / 1,483 ft
*Kuala Lumpur, 1998*

Petronas Tower 2
452 m / 1,483 ft
*Kuala Lumpur, 1998*

Zifeng Tower
450 m / 1,476 ft
*Nanjing, 2010*

Willis Tower
442 m / 1,451 ft
*Chicago, 1974*

Kingkey 100
442 m / 1,449 ft
*Shenzhen, 2011*

Guangzhou IFC
439 m / 1,439 ft
*Guangzhou, 2010*

▲ Diagram of the World's Tallest 20 Buildings according to the CTBUH Height Criteria of "Height to Architectural Top" (as of April 2012).

Although number of floors is a poor indicator of defining a tall building due to the changing floor-to-floor height between differing buildings and functions (e.g., office versus residential usage), a building of perhaps 14 or more stories – or over 50 meters (165 feet) in height – could perhaps be used as a threshold for considering it a "tall building."

## What is a Supertall Building?

The CTBUH defines "supertall" as a building over 300 meters (984 feet) in height. Although great heights are now being achieved with built tall buildings – in excess of 800 meters (2,600 feet) – as of April 2012 there are only approximately 60 buildings in excess of 300 meters completed and occupied globally.

## How is a tall building measured?

The CTBUH recognizes three categories for measuring building height:

▸ **Height to Architectural Top:**
Height is measured from the level[1] of the lowest, significant,[2] open-air,[3] pedestrian[4] entrance to the architectural top of the building, including spires, but not including antennae, signage, flagpoles or other functional-technical equipment.[5] This measurement is the most widely utilized and is employed to define the Council on Tall Buildings and Urban Habitat (CTBUH) rankings of the "World's Tallest Buildings."

▸ **Highest Occupied Floor:**
Height is measured from the level[1] of the lowest, significant,[2] open-air,[3] pedestrian[4] entrance to the finished floor level of the highest occupied[6] floor within the building.

▸ **Height to Tip:**
Height is measured from the level[1] of the lowest, significant,[2] open-air,[3] pedestrian[4] entrance to the highest point of the building, irrespective of material or function of the highest element (i.e., including antennae, flagpoles, signage and other functional-technical equipment).

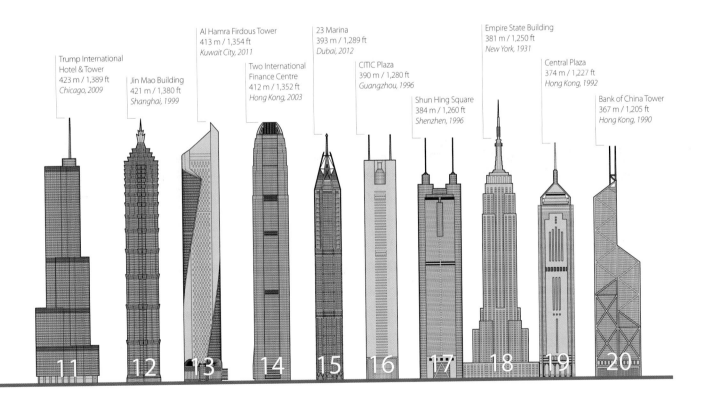

Trump International Hotel & Tower
423 m / 1,389 ft
*Chicago, 2009*

Jin Mao Building
421 m / 1,380 ft
*Shanghai, 1999*

Al Hamra Firdous Tower
413 m / 1,354 ft
*Kuwait City, 2011*

Two International Finance Centre
412 m / 1,352 ft
*Hong Kong, 2003*

23 Marina
393 m / 1,289 ft
*Dubai, 2012*

CITIC Plaza
390 m / 1,280 ft
*Guangzhou, 1996*

Shun Hing Square
384 m / 1,260 ft
*Shenzhen, 1996*

Empire State Building
381 m / 1,250 ft
*New York, 1931*

Central Plaza
374 m / 1,227 ft
*Hong Kong, 1992*

Bank of China Tower
367 m / 1,205 ft
*Hong Kong, 1990*

11  12  13  14  15  16  17  18  19  20

## Number of Floors:

The number of floors should include the ground floor level and be the number of main floors above ground, including any significant mezzanine floors and major mechanical plant floors. Mechanical mezzanines should not be included if they have a significantly smaller floor area than the major floors below. Similarly, mechanical penthouses or plant rooms protruding above the general roof area should not be counted. Note: CTBUH floor counts may differ from published accounts, as it is common in some regions of the world for certain floor levels not to be included (e.g., the level 4, 14, 24, etc. in Hong Kong).

## Building Usage:

What is the difference between a tall building and a telecommunications/observation tower?

▶ A tall "building" can be classed as such (as opposed to a telecommunications/observation tower) and is eligible for the "tallest" lists if at least 50 percent of its height is occupied by usable floor area.

## Single-Function and Mixed-Use Buildings:

▶ A single-function tall building is defined as one where 85 percent or more of its total floor area is dedicated to a single usage.

▶ A mixed-use tall building contains two or more functions (or uses), where each of the functions occupy a significant proportion[7] of the tower's total space. Support areas such as car parks and mechanical plant space do not constitute mixed-use functions.

Functions are denoted on CTBUH "tallest" lists in descending order, e.g., "hotel/office" indicates hotel function above office function.

## Building Status:

▶ **Complete (Completion):**
A building is considered to be "complete" (and officially added to the CTBUH Tallest Buildings lists) if it fulfills all of the following three criteria:
(i) topped out structurally and architecturally,
(ii) fully clad, and
(iii) open for business, or at least occupiable.

▶ **Under Construction (Start of Construction):**
A building is considered to be "under construction" once site clearing has been completed and foundation/piling work has begun.

▶ **Topped Out:**
A building is considered to be "topped out" when it is under construction, and has reached its full height both structurally and architecturally (e.g., including its spires, parapets, etc.).

▶ **Proposed (Proposal):**
A building is considered to be "proposed" (i.e., a real proposal) when it fulfills all of the following criteria:

(i) has a specific site with ownership interests within the building development team,
(ii) has a full professional design team progressing the design beyond the conceptual stage,
(iii) Has obtained, or is in the process of obtaining, formal

planning consent/legal permission for construction, and
(iv) has a full intention to progress the building to construction and completion.

▶ **Vision:**
A building is considered to be a "vision" when it either:
(i) is in the early stages of inception and does not yet fulfill the criteria under the "proposal" category, or
(ii) was a proposal that never advanced to the construction stages, or
(iii) was a theoretical proposition.

▶ **Demolished:**
A building is considered to be "demolished" after it has been destroyed by controlled end-of-life demolition, fire, natural catastrophe, war, terrorist attack, or through other means intended or unintended.

## Structural Material:

▶ A **steel** tall building is defined as one where the main vertical and lateral structural elements and floor systems are constructed from steel.

▶ A **concrete** tall building is defined as one where the main vertical and lateral structural elements and floor systems are constructed from concrete.

▶ A **composite** tall building utilizes a combination of both steel and concrete acting compositely in the main structural elements, thus including a steel building with a concrete core.

▶ A **mixed-structure** tall building is any building that utilizes distinct steel or concrete systems above or below each other. There are two main types of mixed structural systems: a **steel/concrete** tall building indicates a steel structural system located above a concrete structural system, with the opposite true of a **concrete/steel** building.

▸ **Additional Notes on Structure:**
(i) If a tall building is of steel construction with a floor system of concrete planks on steel beams, it is considered a **steel** tall building.
(ii) If a tall building is of steel construction with a floor system of a concrete slab on steel beams, it is considered a **steel** tall building.
(iii) If a tall building has steel columns plus a floor system of concrete beams, it is considered a **composite** tall building.

---

[1] Level: finished floor level at threshold of the lowest entrance door.

[2] Significant: the entrance should be predominantly above existing or pre-existing grade and permit access to one or more primary uses in the building via elevators, as opposed to ground floor retail or other uses which solely relate/connect to the immediately adjacent external environment. Thus entrances via below-grade sunken plazas or similar are not generally recognized. Also note that access to car park and/or ancillary/support areas are not considered significant entrances.

[3] Open-air: the entrance must be located directly off of an external space at that level that is open to air.

[4] Pedestrian: refers to common building users or occupants and is intended to exclude service, ancillary, or similar areas.

[5] Functional-technical equipment: this is intended to recognize that functional-technical equipment is subject to removal/addition/change as per prevalent technologies, as is often seen in tall buildings (e.g., antennae, signage, wind turbines, etc. are periodically added, shortened, lengthened, removed and/or replaced).

[6] Highest occupied floor: this is intended to recognize conditioned space which is designed to be safely and legally occupied by residents, workers or other building users on a consistent basis. It does not include service or mechanical areas which experience occasional maintenance access, etc.

[7] This "significant proportion" can be judged as 15 percent or greater of either: (i) the total floor area, or (ii) the total building height, in terms of number of floors occupied for the function. However, care should be taken in the case of supertall towers. For example a 20-story hotel function as part of a 150-story tower does not comply with the 15 percent rule, though this would clearly constitute mixed-use.

# CTBUH Organization and Members

## CTBUH Organizational Members
(as of April 2012)

*http://membership.ctbuh.org*

### Supporting Contributors
AECOM
Al Hamra Real Estate Company
NV. Besix SA
BT – Applied Technology
Buro Happold, Ltd.
Daewoo Engineering & Construction Co., Ltd.
Emaar Properties, PJSC
HOK, Inc.
Hyundai Amco Co., Ltd.
Hyundai Development Company
Illinois Institute of Technology
Kohn Pedersen Fox Associates, PC
KONE Industrial, Ltd.
Lotte Engineering & Construction Co.
NBBJ
Samsung C&T Corporation
Shanghai Tower Construction & Development Co., Ltd.
Shree Ram Urban Infrastructure, Ltd.
Skidmore, Owings & Merrill LLP
Taipei Financial Center Corp. (TAIPEI 101)
Turner Construction Company
Woods Bagot

### Patrons
Arabtec Construction LLC
Blume Foundation
BMT Fluid Mechanics, Ltd.
The Durst Organization
East China Architectural Design & Research Institute Co., Ltd.
FC Beekman Associates, LLC
Gensler
Hongkong Land, Ltd.
KLCC Property Holdings Berhad
Kuwait Foundation for the Advancement of Sciences (KFAS)
Meinhardt Group International
Studio Daniel Libeskind
Thornton Tomasetti, Inc.
Tishman Speyer Properties
Weidlinger Associates, Inc.
Zuhair Fayez Partnership

### Donors
Adrian Smith + Gordon Gill Architecture, LLP
American Institute of Steel Construction
Aon Fire Protection Engineering Corp.
Arup
Aurecon
Bollinger + Grohmann Ingenieure
C.Y. Lee & Partners Architects/Planners
CCDI (China Construction Design International)
Enclos Corp.
Fender Katsalidis
Gale International / New Songdo International City Development , LLC
Halfen USA
Heerim Architects & Planners Co., Ltd.
Hyundai Steel Company
Jacobs
Larsen & Toubro, Ltd.
Leslie E. Robertson Associates, RLLP

Magnusson Klemencic Associates, Inc.
Mooyoung Architects & Engineers
Paragon International Insurance Brokers Ltd.
Pei Cobb Freed & Partners
Pickard Chilton
Rafik El-Khoury & Partners
Remaking of Mumbai Federation
The Rise Group LLC
RMJM
Rolf Jensen & Associates, Inc.
Rowan Williams Davies & Irwin, Inc.
RTKL
Severud Associates Consulting Engineers, PC
Shanghai Construction (Group) General Co. Ltd.
Shanghai Institute of Architectural Design & Research Co., Ltd.
SIAPLAN Architects and Planners
Solomon Cordwell Buenz
SsangYong Engineering & Construction Co., Ltd.
Studio Gang Architects
SWA Group
Syska Hennessy Group, Inc.
Viracon
Walter P. Moore and Associates, Inc.
Werner Voss + Partner
Yolles

### Contributors
Aedas, Ltd.
ALHOSN University
Alvine Engineering
American Iron and Steel Institute
Ayala Land, Inc.
Barker Mohandas, LLC
Bates Smart
Benoy Limited
Bonacci Group
Boundary Layer Wind Tunnel Laboratory
The British Land Company PLC
Broadway Malyan Asia Pte Ltd
C.S. Structural Engineering, Inc.
Canary Wharf Group, PLC
Canderel Management, Inc.
CCL Qatar w.l.l.
Continental Automated Buildings Association (CABA)
Daelim Industrial Co., Ltd.
Davis Langdon & Seah Singapore Pte. Ltd.
DBI Design Pty Ltd
DCA Architects Pte Ltd
Deerns Consulting Engineers
DHV Bouw en Industrie
DongYang Structural Engineers Co., Ltd.
Dow Corning Corporation
EW Cox Hong Kong
Far East Aluminum Works (US) Corporation
Goettsch Partners
HAEAHN Architecture, Inc.
Hiranandani Group
International Paint Ltd.
JCE Structural Engineering Group, Inc.
Jiang Architects & Engineers
KHP Konig und Heunisch Planungsgesellschaft
Lend Lease
M Moser Associates Ltd.
Mitsubishi Electric Corporation
Mori Building Co., Ltd.
MulvannyG2 Architecture
Mutua Madrilena
Nabih Youssef & Associates
National Fire Protection Association
National Institute of Standards and Technology (NIST)
Nishkian Menninger Consulting and Structural Engineers
Norman Disney & Young
Otis Elevator Company
Paris La Défense (Etablissement Public d'Aménagement de La Défense Seine Arche)
Perkins + Will
Permasteelisa North America
PositivEnergy Practice, LLC

RAW Design Inc.
Rosenwasser/Grossman Consulting Engineers, PC
SAMOO Architects & Engineers
Sanni, Ojo & Partners
Schindler Elevator Corporation
Silvercup Studios
SilverEdge Systems Software, Inc.
SIP Project Managers Pty Ltd
The Steel Institute of New York
T. R. Hamzah & Yeang Sdn. Bhd.
Tekla Corp.
ThyssenKrupp Elevator Qatar
TSNIIEP for Residential and Public Buildings
University of Illinois at Urbana-Champaign
wh-p GmbH Beratende Ingenieure
Wilkinson Eyre Architects
WSP Group

## Participants

ACSI (Ayling Consulting Services Inc)
Aidea Philippines, Inc.
AKF Group, LLC
Al Ghurair Construction – Aluminum LLC
Al Jazera Consultants
ALT Cladding, Inc.
ARC Studio Architecture + Urbanism
ArcelorMittal
Architects 61 Pte., Ltd.
Architectural Design & Research Institute of Tsinghua
    University
Architectural Institute of Korea
Arquitectonica International Corp.
Atkins
B+H Architects
Bakkala Consulting Engineers Limited
BAUM Architects, Engineers & Consultants, Inc.
BDSP Partnership
Beazley Group Plc
Beca Group
BFLS
BG&E Pty., Ltd.
Bigen Africa Services (Pty) Ltd.
Billings Design Associates, Ltd.
BluEnt
Boston Properties, Inc.
Bouygues Construction
Callison, LLC
Capital Group
Case Foundation Co.
CB Engineers
CCHRB (Chicago Committee on High-Rise Buildings)
CDC Curtain Wall Design & Consulting, Inc.
China Academy of Building Research
Chinachem Group
CICO Consulting Architects and Engineers
City Developments Limited
Clean Urban Energy, Inc.
Code Consultants, Inc.
Cook+Fox Architects
Cosentini Associates
COWI A/S
Cox Architecture Pty. Ltd.
CPP, Inc.
CS Associates, Inc.
CTL Group
Cundall
Dam & Partners Architecten
Dar Al-Handasah (Shair & Partners)
Delft University of Technology
Dennis Lau & Ng Chun Man Architects & Engineers (HK), Ltd.
dhk Architects Pty., Ltd.
DSP Design Associates Pvt., Ltd.
Dunbar & Boardman
Edgett Williams Consulting Group, Inc.
EEI Corporation
Electra Construction LTD
ELU Konsult AB

Ennead Architects LLP
Environmental Systems Design, Inc.
Epstein
Eser Contracting & Industry Co.
Fortune Consultants, Ltd.
FXFOWLE Architects, LLP
M/s. Glass Wall Systems (India) Pvt. Ltd
Gold Coast City Council
Gorproject (Urban Planning Institute of Residential and Public
    Buildings)
Grace Construction Products
Greyling Insurance Brokerage
Grupo Elipse
Guangzhou Scientific Computing Consultants Co., Ltd.
GVK Elevator Consulting Services, Inc.
Halvorson and Partners
Haynes-Whaley Associates, Inc.
Heller Manus Architects
Hilson Moran Partnership, Ltd.
Hong Kong Housing Authority
BSE, The Hong Kong Polytechnic University
Housing and Development Board
IECA Internacional S.A.
Infrastrutture Lombarde S.p.A.
Institute BelNIIS, RUE
INTEMAC, SA
Irwinconsult Pty., Ltd.
Iv-Consult b.v.
Jaros Baum & Bolles
JBA Consulting Engineers, Inc.
JMB Realty Corporation
John Portman & Associates, Inc.
JV "Alexandrov-Passage" LLC
Kajima Overseas Asia Pte Ltd
Kalpataru Limited
KEO International Consultants
Kinetica Dynamics Inc.
King Saud University College of Architecture & Planning
Korea University
The Korean Structural Engineers Association
KPMB Architects
Langan Engineering & Environmental Services, Inc.
Leigh & Orange, Ltd.
Lerch Bates, Inc.
Lerch Bates, Ltd. Europe
Lobby Agency
Louie International Structural Engineers
Mace Group
Magellan Development Group, LLC
Margolin Bros. Engineering & Consulting, Ltd.
James McHugh Construction Co.
McNamara/Salvia, Inc.
Mesa Development, LLC
Metropro Consultants India Pvt. Ltd.
Murphy/Jahn Architects LLC
Nanjing International Group Co. Ltd.
Nikken Sekkei, Ltd.
O'Connor Sutton Cronin
Odell Associates, Inc.
Option One International, WLL
P&T Group
Palafox Associates
PDW Architects
Pelli Clarke Pelli Architects
Perkins Eastman Architects, PC
Powe Architects
PPG Industries, Inc.
Project and Design Research Institute "Novosibirsky
    Promstroyproject"
Rafael Vinoly Architects, PC
Read Jones Christoffersen Ltd.
Rene Lagos Engineers
Riggio / Boron, Ltd.
RMIT University
Rodium Properties
Ronald Lu & Partners
Roosevelt University – Marshall Bennett Institute of Real
    Estate

S.K.S. Associates
Sauerbruch Hutton Verwaltungsges mbH
schlaich bergermann und partner
Sematic SPA
Shimizu Corporation
Siemens Industry
Sinosteel Ever Glory Co., Ltd.
SmithGroup
SOMA Architects
St. Francis Square Development Corp.
Stanley D. Lindsey & Associates, Ltd.
Stephan Reinke Architects, Ltd.
Studio Altieri S.p.A.
Taisei Corporation
Takenaka Corporation
Tameer Holding Investment LLC
Tandem Architects (2001) Co., Ltd.
Taylor Thomson Whitting Pty., Ltd.
TFP Farrells, Ltd.
Thermafiber, Inc.
Tongji Architectural Design (Group) Co., Ltd.
Transsolar
The Trump Organization
United Elevator Consultants, Inc.
University of Maryland – Architecture Library
University of Nottingham
UralNIIProject RAACS
Vanguard Realty Pvt., Ltd.
VDA (Van Deusen & Associates)
Vipac Engineers & Scientists, Ltd.
VOA Associates, Inc.
Walsh Construction Company
Werner Sobek Stuttgart GmbH & Co., KG
Windtech Consultants Pty., Ltd.
WOHA Architects Pte., Ltd.
Wong & Ouyang (HK), Ltd.
Wordsearch
World Academy of Science for Complex Safety
WSP Cantor Seinuk
WSP Flack + Kurtz, Inc.
WTM Engineers International GmbH
WZMH Architects
Y. A. Yashar Architects

*Supporting Contributors are those who contribute
$10,000; Patrons: $6,000; Donors: $3,000;
Contributors: $1,500; Participants: $750.*

697.92 .W66 2013
Natural Ventilation
in High-Rise
Office Buildings

## Date Due